Image Processing With Xilinx Devices

Adam P Taylor

Copyright © 2017 Adam Taylor

All rights reserved.

ISBN: 1981943285
ISBN-13: 978-1981943289

DEDICATION

To my son Daniel Peter Steven Taylor

CONTENTS

	Acknowledgments	i
1	INTRODUCTION TO IMAGE PROCESSING	1
2	GETTING DOWN WITH EMBEDDED VISION ALGORITHMS	18
3	BUILDING BLOCKS OF AN EMBEDDED VISION SYSTEM	24
4	USING A ZYNQ WITHIN YOUR EMBEDDED VISION SYSTEM	36
5	USING A FPGA WITHIN YOUR EMBEDDED VISION SYSTEM	45
6	INTERFACING: CAMERA LINK	55
7	INTERFACING: HDMI	60
8	USING HIGH LEVEL SYNTHESIS	77
9	TRANSFERING IMAGES FROM THE PL TO PS	103
10	ADDRESSING VDMA ISSUES	108
11	VERIFYING VIDEO IMAGES IN ZYNQ SOC AND ZYNQ ULTRASCALE+ MPSOC MEMORY	111
12	Video Mixing	121
13	Synchronising Output video	127
14	TIPS FOR BETTER IMAGE PROCESSING SYSTEMS	131
15	FLIR Lepton Example	137
16	Example Designs on the Web	147

ACKNOWLEDGMENTS

I would like to thank all those who read the blogs weekly, but especially to Steve Leibson who started it all.

INTRODUCTION TO IMAGE PROCESSING

Vision based systems have become commonplace across many industries and applications, indeed many of us carry an embedded vision system daily in our smartphones. These devices are capable of not only capturing images and recording video but also performing augmented reality applications demonstrating just how accepted embedded vision has become.

The growth of embedded vision applications across both traditional and emerging applications (examples of which are shown in Figure 1) have arisen as a result of increases in processing power, memory density and tighter system integration. This has allowed embedded vision applications to become widely accepted by both consumer, industry and governments, acceptance which will lead to significant growth before the decade is out. Table 1 identifies some predicted high growth areas for embedded vision applications some of which are obvious and others which are not quite.

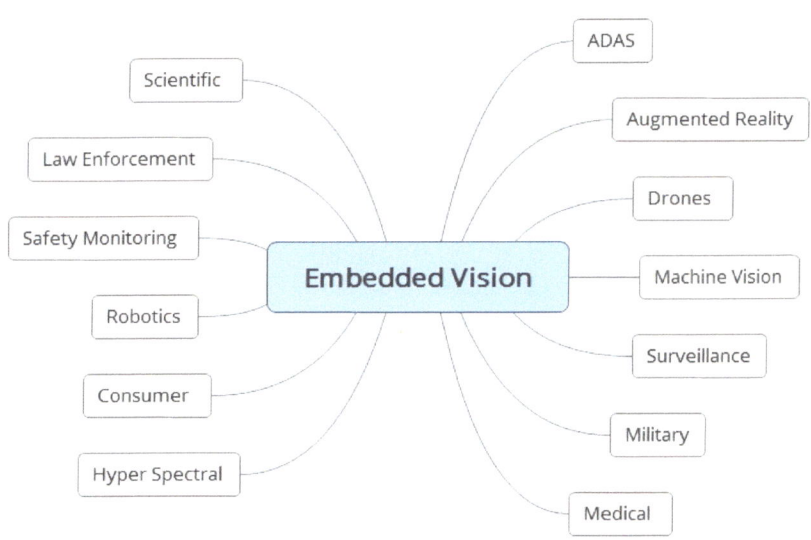

Figure 1 Common Embedded Vision Applications.

Application	Example Sector	CAGR[1]	Year
Machine Vision	Machine Vision	9.1 %	2020
Hyper Spectral Imaging	Hyper Spectral	12%	2019
Toll Collection	Surveillance	11.1%	2020
3 Dimensional Imaging	ADAS, Medical,	27.5%	2020
Facial Recognition	Surveillance	17.4%	2020

Table 1 predicted high growth rate areas for embedded vision applications

But what is embedded vision?

An embedded vision system comprises the entire signal chain from reception of the photon by the selected imaging sensor to the system output, be that a processed / unprocessed images or information extracted from the image and provided to a downstream system. Of course the embedded system architect is responsible for guaranteeing the end to end performance against the system requirements.

Achieving this requires the embedded vision system architect to be familiar with a range of concepts and techniques, both related to the senor and post processing system. This paper is intended to act as a high level primer to provide a basic understanding of these techniques and concepts.

The first thing we must be familiar with is the electromagnetic spectrum and the area of the spectrum which we wish our system to operate. While humans have a limited ability to see only the range of wavelengths between the Blue (390 nm) and Red (700nm) section of the spectrum, commonly referred to as the visible spectrum, imaging

---

[1] CAGR – Compounded Annual Growth Rate, this is the year over year growth rate.

Image Processing With Xilinx Devices

devices depending upon the technology chosen, can image across a much wider range including X-ray, UV and Infra-Red in addition to the visible spectrum.

Working within the Near IR Spectrum and below we can use devices such as Charge Coupled Devices (CCDs) or CMOS[2] (Complementary Metal Oxide Semiconductor) Image Sensors (CIS) as we move into the infrared spectrum we need to use specialized IR detectors. The need for specialized sensors in the IR domain is in one part due to the excitation energy required for silicon based imagers such as CCD or CIS. These typically require photon energy of 1eV to excite an electron however within the IR domain photon energies range from 1.7eV to 1.24 meV as such IR imagers tend to be based upon HgCdTe or InSb. These have lower excitation energies and are often combined with a CMOS readout IC called a ROIC to control and readout the sensor.

The two most common detector technologies are CCD or CIS

- Charge Coupled Devices are best considered as analog devices, as such their integration into a digital system requires off device ADCs and clock generation at the required analog voltage levels. Each pixel stores the charge generated by the photon, for most applications the pixels are arranged in a 2 dimensional array arranged as a number of lines, where each line contains a number of pixels. Read out of the CCD uses line transfer to transfer each of the lines in parallel down into a read out register where each line is read out serially. It is in this register readout process that the charge is converted to a voltage from the charge.

- CMOS Imaging Sensors allow much tighter integration with ADCs, Bias and Drive circuits to be integrated on the same die. This significantly decreases system integration while at the same time increases the complexity of the CIS design. The heart

---

[2] We can use different coatings upon the image device to affect its wavelength performance.

of the CIS is the Active Pixel Sensor (APS) which includes both the photodiode and the readout amplifier in each pixel unlike CCDs this allows a CIS to readout address any pixel in the array.

While the majority of embedded vision applications use CIS devices, CCD's are still used within high end scientific applications where performance is critical. The elements of the paper to follow are applicable to both imaging technologies.

Sensor Considerations

Selecting the correct sensor for our application requires understanding the system requirements however there are several aspects of the device we must consider to help us achieve these.

The first requirement is that we must determine the resolution required, that is how many pixels should be present on each line and how many lines the detector requires. The end application will drive this significantly, as an example a scientific astronomy application may have a requirement for a high resolution 2 dimensional device while an industrial inspection imaging application may require only a line scan approach.

- Line scan devices consist of a single line (sometimes a few lines) of pixels in the X direction, these are used typically for applications where either the camera or the target is moving allowing the image in the Y direction to be generated. As such they are used for inspection applications or Optical Character Recognition (OCR) applications. Some applications require the use of Time Domain Integration (TDI) line scan sensors these consist of multiple lines in the X direction as the target moves, so does the pixel value from one to the next allowing for a more sensitive detection as charge is integrated over time. TDI however requires synchronization between the line transfer and movement of the target, to prevent smearing and image defects. As there are only a few lines to read out frame rates can be very high.

Image Processing With Xilinx Devices

- Two Dimensional array these consist of a number of lines each with a number of pixels, the size of the array will be one factor which determines the maximum frame rate of the sensor. Typically to achieve higher frame rates two dimensional devices readout and in a parallel number of pixels. It is also possible with 2D devices to perform what is called windowing (sometimes called Region of Interest) and only read out a particular region of interest in the device to obtain a higher frame rate. These devices are used in many applications where the information is contained within the 2 dimensional images for instance advanced driver assistance systems (ADAS), surveillance or scientific applications.

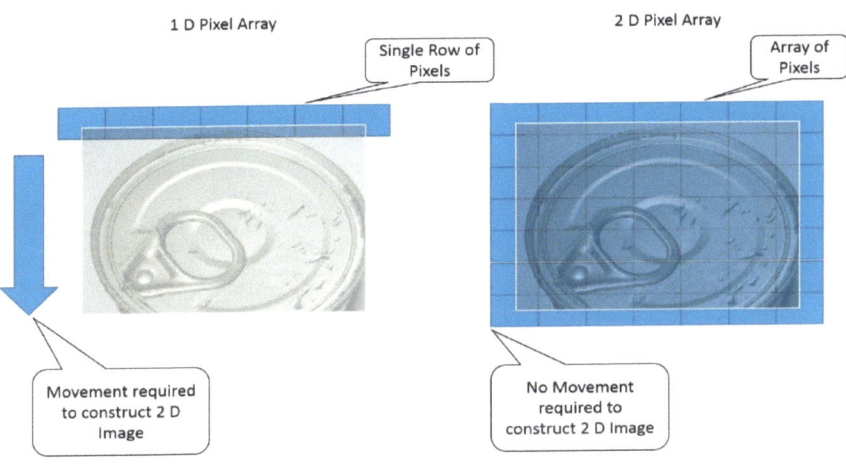

Figure 2 Line Scan and Array Sensor concept

Having determined the format of the imager and the resolution required we must also consider the pitch of the pixels. The pitch defines the size of the pixel which is available to collect charge created by incident photons. As such a smaller pixel pitch will mean that during the integration period (the time the sensor is exposed to the image) a smaller charge is collected. With a smaller pixel pitch this can mean that longer integration times are required to capture the image, this can impact the ability of the sensor to image fast moving images and reduce its low light performance.

Once the sensor format has been determined we must then consider the technology CCD, CMOS or more specialized. The key driving parameter for this is the Quantum Efficiency (QE) at its most basic this is the efficiency of the device at producing electrons for photons striking the device. Typically, we want to achieve as high a QE as possible across the spectrum of interest, this is also of great importance when performing low light applications. There are three driving aspects which can affect QE of a device absorption, reflection and transmission. One large contributor to the degradation of QE is the structure of the device which may result in the pixel being shielded by circuitry within the sensor for example metallic lines or poly silicon gates etc. These structures absorb or reflect the photons degrading the QE this leads to the choice in sensors

- Front illuminated – In these devices the photons strike the front face of the device in the traditional manner explained above, that is the pixel may be obscured and QE is reduced accordingly
- Back Illuminated – These devices under take post processing to thin the back of the device such that the illumination is received on the back and there are no obstructions from other design elements. The best QE is obtained with back thinned devices.

Image Processing With Xilinx Devices

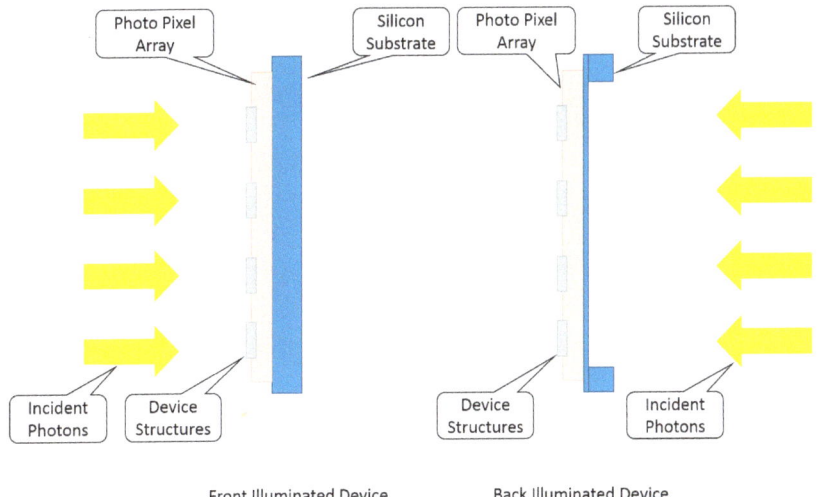

Figure 3 Front and Back Illuminated Concept

We must also consider the noise allowable within the image sensor; there are three main areas we must consider.

- Device Noise – This is temporal in nature and includes shot noise, noise introduced by the output amplifiers and the reset circuits.
- Fixed Pattern Noise (FPN) – Is spatial in nature and caused by the different responses of pixels when subject to the same illumination intensity. FPN is normally caused by differing offset and gain response from each pixel the offset is often called the Dark Signal Non Uniformity (DSNU) while the gain is given by the Photo Response Non Uniformity (PRNU). There are a number of techniques to compensate for FPN however one of the most popular is correlated double sampling of the output signal.
- Dark Current - This is caused by thermal noise within the image sensor and is present even in the case of no illumination. The impact of dark signal upon the final image quality will depend upon the frame rate, at higher frame rates it does not dominate however as frame rates decrease as say for scientific applications then it can dominate. As dark current is

temperature related for applications which require a reduced dark current it is typical to reduce the operating temperature of the imaging device using a cooling device like Peltier.

Understanding the noise model of the imager allows us to determine what Signal to Noise Ratio (SNR) is achievable.

Having determined the noise performance of the device allows us to determine the dynamic range required by the image sensor. Dynamic Range represents the ability of the sensor to capture images which contains illumination intensities over a wide range this is usually given in dB or as a ratio. This means the image can contain both high illumination areas and dark areas in the same image.

The actual Dynamic Range of the sensor is determined by the full well capacity of the pixel, that is the number of electrons the pixel can hold before it saturates. Expressed as a ratio against the readout noise in electrons, this ratio can then be converted to dB with ease.

$$\frac{Full\ Well\ Capacity\ e-}{Readout\ Noise\ e-}$$

The dynamic range is often determined by performing a Photon Transfer Curve test; this plots noise against the well capacity.

Of course if the device has a digital output this relates directly to the number of bits provided on the output using the formula below

$$Dynamic\ Range\ (dB) = 20\ Log10(2^{Bus\ Width})$$

This does not however guarantee that the device will achieve the dynamic range; it just indicates its potential range which can be represented by the bus width regardless of sensor performance.

The IO standard used to output not only the pixel's data but also the command and control interface will also be of importance. This will correlate to the frame rate for instance a LVCMOS interface is not suitable for a high frame rate application, but is acceptable for a simple monitoring camera. The trend for imaging devices is towards dedicated high speed serialised links which use the LVDS family or SERDES technology as frame rates, resolution and bits per pixel increase.

So far we have explored many different important aspects of the image sensor however once aspect we have not yet considered is if the imager is to be a color or monochrome sensor. Both selections depend upon the application as would be expected

- Color Sensor – Requires the use of a Bayer pattern on top of each pixel alternating red, green on one line and blue, green on the next (There is more green as the human eye is more sensitive to green wavelengths). This means each pixel only receives photons in the required wavelength filtering the photons received. We can determine the color of the pixel by post processing the image to reconstruct the color at each pixel using the different colors surrounding pixels to determine the color although this does reduce the resolution of the image. Using a color sensor also complicates the image processing chain required to reconstruct and output an image. The use of a Bayer pattern does result in a lower resolution although not as much as might first be thought, the resolution is typically degraded by 20%.
- Monochrome – As there is no Bayer pattern on top of the image array each pixel receives all of the photons. This allows for an increase in sensitivity of the image and allows for simpler readout of the image as there is no de-mosaicking required for the color reconstruction.

Should our sensor selection conclude a CIS device is to be used, these are in reality complex dedicated system on chips. As such we must also take into account the following aspects related to readout modes and integration time.

- Integration time – This is the time the pixels are exposed to illumination before readout, on simpler CCD based systems the proximity electronics will perform the timing external to the device. However, with CIS devices the integration time will be configured via a register over the command interface and then the CIS device will perform the integration time accurately for one of two commonly used readout modes.

- Global Shutter Mode – In this mode all of the pixels are exposed to illumination at once then then read out, in this mode as all pixels are read out read noise will be increased however this mode is a good mode to use if we want to snap a shot of a fast moving target.
- Rolling Shutter Mode – In this mode each line is exposed to illumination and read out, in this mode the read noise is reduced, however it is not as suitable as global shutter mode for capturing a fast moving target.

System development

Once we have selected the correct sensor for our application there are still a number of challenges and decisions which must be considered and addressed for the system development.

Image Processing With Xilinx Devices

Figure 4 Embedded Vision Development System

Along with the technical challenges, development of the system will face time pressures, to ensure the desired time to market is achieved. We can use this time constraint to focus upon understanding what our value added activity is in the development and making the correct choices with regards to what is developed (the value added activity) and what is purchased as commercial off-the-shelf (COTS) or subcontracted for development. Focusing on the value added activates and leveraging IP modules at the hardware, SW and FPGA levels is one of the key enabling factors to meeting time to market.

Along with time to market embedded vision systems are also often developed for applications where size, weight and power often called SWaP-C where the 'C' represents cost are also driving factors. What dominates in the SWaP-C of your system will depend upon the application, for instance a handheld unit will have more stringent requirements on power than an ADAS application. However, for an ADAS application the cost of the solution will be a driving factor as many millions of units will be produced.

One way to achieve better SWaP-C performance is through tighter system integration both at the sensor and in the processing system that is fewer, but more capable integrated components.

While each application will have different areas of added value and SWaP-C considerations almost all embedded vision systems will require that we implement what is commonly referred to as the image processing pipeline. This pipeline will interface with the sensor selected and perform the operations required to produce an image suitable for either further processing or transmission over a network. A basic image processing pipeline may consist of

- Camera interface – Reception to the RAW image from the senor
- Color Filter Array – Reconstruction of the pixel colors
- Color Space Conversion – Conversion to the correct color space for Codec etc.
- Output formatting - Interfacing with the output medium

It is within this image processing pipeline that we will perform and apply our algorithms to the received images. Depending upon the application being implemented these algorithms will vary however, there are a number of commonly used image processing algorithms which can be used to improve the contrast, detect features, objects or movement in an image or correct for blurring of the image.

These algorithms should be developed within a framework which enables us to get our product to market in the fastest possible time and encourage reuse while reducing non-recurring and recurring

Image Processing With Xilinx Devices

engineering cost (NRE & RE). There exist several potential frameworks which we should consider.

- OpenVX – Open source application for development of image processing applications
- OpenCV[3] – Open Source Computer Vision a number of libraries aimed at real time computer vision based on C / C++
- OpenCL – Open Source Computer Language based upon C++ for developing applications for parallel processed applications as seen in GPU, FPGA, etc.
- SDSoC – A design environment from Xilinx that allows developers to initially implement algorithms written in C / C++ in the ARM processing system of a Zynq or UltraScale+ MPSoC device, profile the code base to identify performance bottlenecks, and then using Xilinx High Level Synthesis to translate those bottlenecks into hardware-enabled IP that run in the programmable logic (PL) portion of the device.

Use of these frameworks coupled with HLS if we follow a FPGA or All Programmable SoC approach allows for efficient development of embedded vision applications which can be quickly demonstrated with hardware in the loop.

Once the image completes the processing pipeline how the data is output from the system is also important, at the highest level there are three high level choices

- The image is output on to a display using standards like Video Graphics Array (VGA), High Definition Multimedia Interface (HDMI), Serial Digital Interface (SDI) or DisplayPort. In many EV applications displays may also be touch screen enabled to control and configure the system.

---

[3] See XAPP 1167
http://www.xilinx.com/support/documentation/application_notes/xapp1167.pdf

- The image or information extracted from the image, is transmitted to another system which uses the image or image extracted as would be the case in cloud processed applications
- The images are stored on non-volatile media to be accessed at a later date.

For the majority of these high level choices at the completion of the imaging chain we will need to determine how the image is formatted for use. This presents us with the choice of if we wish to encode the image using an industry standard image / video compression algorithm such as H.264 (MPEG-4 Part 10 Advanced Video Coding) or H.265 (High Efficiency Video Coding) implementations of these are often called Codec. Codecs allow for more efficient utilization of communication and network bandwidth or a reduction in the storage footprint required for a small impact in loss of fidelity as the encoding is generally lossy[4]. Alternatively, if loss of fidelity is not acceptable due to the application then the image can be transmitted or stored in its raw format or encoded in a lossless format.

Most codec implementations use a different color space to that which is output by the image sensor, should the system use a color device. The mainly used color spaces within embedded vision are:

- Red, Green, Blue – This contains the RGB information as output from the image sensor, it is commonly used as an output for simple interfaces like VGA
- YUV – This contains Luma (Y) and the chrominance (U & V) this color space is used for most codecs and some display standards. Commonly used formats of YUV are YUV4:4:4 and YUV4:2:2. The difference between the two formats is that with 4:4:4 each pixel is represented by eight bits making for a 24-bit pixel. With a 4:2:2 format the U & V values are shared between pixels allowing for a 16-bit pixel which proves more memory efficient.

One further decision which will have a considerable impact on the

---

[4] It is possible to create lossless encoded applications

Image Processing With Xilinx Devices

image processing chain and the SWAP-C is where the majority of the image processing chain is implemented:

- At the edge, i.e. within the embedded vision system itself this will drive up the power and processing / memory capabilities of the system required but also will enable a faster response times. Processing at the edge will be the dominant application for most embedded applications like ADAS, machine vision, etc.
- Within the Cloud this requires that the embedded vision system captures the image and is capable of transmitting it to the cloud using network enabled technology. Typical applications which process in the cloud include some medical imaging or scientific applications where the processing can be very intensive and real time results are not required.

To implement the processing chain, the heart of an embedded vision system requires a processing core which is capable of not only controlling the selected image sensor but also receiving, implementing the image processing pipeline and transmitting the images over the desired network infrastructure, or to the chosen display. These demanding requirements often result in a selection of a FPGA or as in more and more cases an All Programmable System on a Chip like the Zynq or Zynq MPSoC devices.

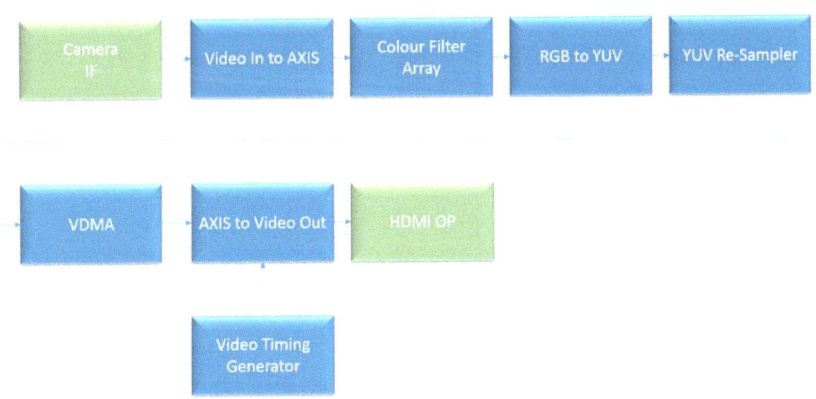

Figure 5 Example Image Processing Chain

The Zynq device combines two high performance ARM processors with

FPGA fabric. This split enables the Processor System (PS) to be used to communicate with the host systems over GigE, PCIe or other interfaces like CAN while performing general management / housekeeping of the system. The programmable logic (PL) half of the device can be used to receive and process the images, exploiting the parallel nature of FPGA fabric. If the images are required to be transmitted over network infrastructure, then Direct Memory Access (DMA) controllers within the Zynq can be used to efficiently move image data from the PL to the PS DDR Memory. Once within the PS DDR memory it can then be further accessed using DMA controllers of the transport medium selected.

Of course once the image is within the PS DDR the high performance processors are also capable of providing further processing operations. The Zynq architecture is such that is it possible to also move processed images from the PS DDR back into the image pipeline in the PL.

Sensor Fusion

Many embedded vision systems also require the ability to integrate additional sensor data to better perceive the environment. This can include the use of many sensors of the same kind (homogeneous sensor fusion) to increase field of view, e.g. like a surround view use case in an ADAS application, or integrate many sensors of varying kinds (heterogeneous sensor fusion) to provide perception into aspects of vision than may not be seen in the visible light spectrum, e.g. overlaying infrared information over regular image sensor data.

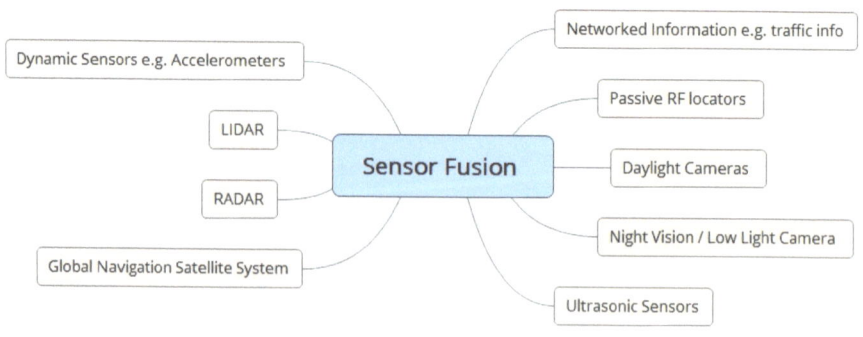

Figure 6 Image Fusion Sources

Image Processing With Xilinx Devices

In many applications the outputs from your embedded vision application will be fused with other sensor data to create an image which contains information from several different sensors. The simplest sensor fusion applications are the combination of images across different spectrums for instance visible fused with IR to enable better night vision during nocturnal hours.

A more complex sensor fusion application would be fusion of the imaging system, Global Navigation Satellite Systems (GNSS), digital mapping information along with other sensors operating in different wavelengths, e.g. RADAR to determine the exact position of another automobile's relative position to enable collision avoidance features.

Sensor fusion can be very processor intensive as the disparate systems are fused together and the information extracted. Again here an all Programmable System on Chip solution offers several significant advantages due to the ability to interface and process several sensors information in parallel increasing the data throughput.

GETTING DOWN WITH EMBEDDED VISION ALGORITHMS

Embedded vision covers a very wide area, so I want to take some time in this blog to look at some other embedded-vision applications and how we can implement the relevant algorithms. Embedded-vision algorithms break down into the following high-level categories:

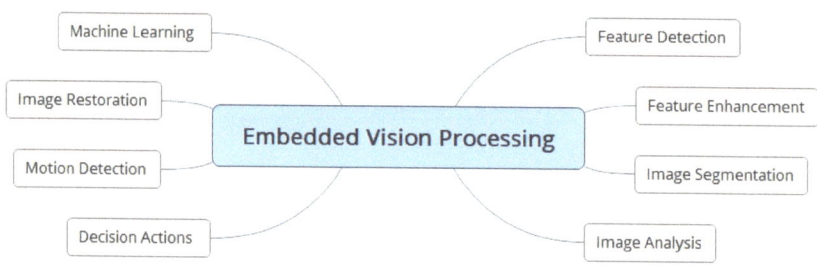

Figure 7 Embedded Vision Categories

These processing techniques split further into tasks that process and extract information from the image and tasks that use the results of these operations for analysis and decision making.

One of the most commonly used embedded-vision techniques is applying filters to the image. Because information within an image resides in the spatial domain and not the frequency domain, image-processing filters typically use convolution filters. That is, we convolve the image with a 2D filter kernel to obtain the desired response.

As with 1D convolution, we must consider the filter's impulse response. This is called the point-spread function (PSF) in image-processing applications. To control the filter's function, we define a custom PSF for each function just as we would define different impulse responses for signal-processing filters (e.g. high-pass, low-pass, etc).

Most commonly, we implement the filter kernel as a 2D matrix, which

Image Processing With Xilinx Devices

we apply to each pixel in the image. Within the implementation, we need to buffer lines so that we can slide the filter across the image.

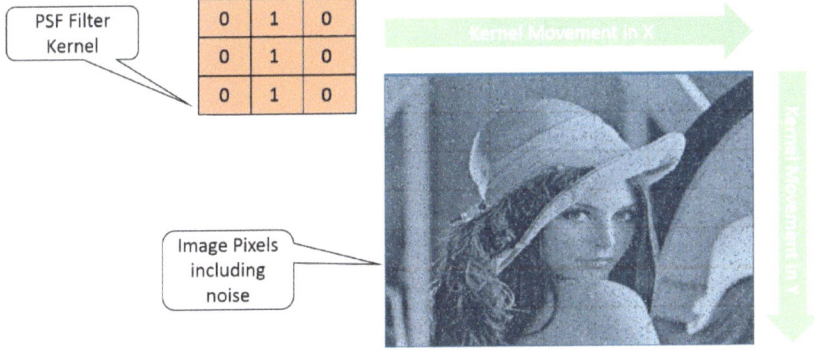

Figure 8 Example Image and filter kernel applied on pixel-by-pixel basis on the Lena image

Within these filter kernels, we can define PSF's for:
- Noise Reduction
- Edge Enhancement
- Edge Detection

These Imaging filters are defined as linear filters. There is another class of filters called non-linear filters that include filters like the median-order filter.

One of the major differences between linear and non-linear filters is their edge-preservation capability. Linear filters can produce blurred edges. Non-linear filters tend to preserve edges.

Figure 9 Example of original Lena image (left) and edge enhancement (Right)

Depending upon the image, we may need to adjust the contrast to extract the most information from in the image. Each pixel has a value within an image. These values will be close together in a low-contrast image, which makes it hard to distinguish between pixels. Contrast enhancement widens the distribution of pixel values to make subsequent image-processing algorithms easier to implement.

The contrast of an image can be determined by its histogram, which shows the distribution of pixel values. A low-contrast image will demonstrate a tight grouping of pixel values in the histogram. A high-contrast image will show a wide spread of values. Commonly used algorithms for contrast enhancement include contrast stretching, which can use linear or non-linear approaches or histogram equalization

Image Processing With Xilinx Devices

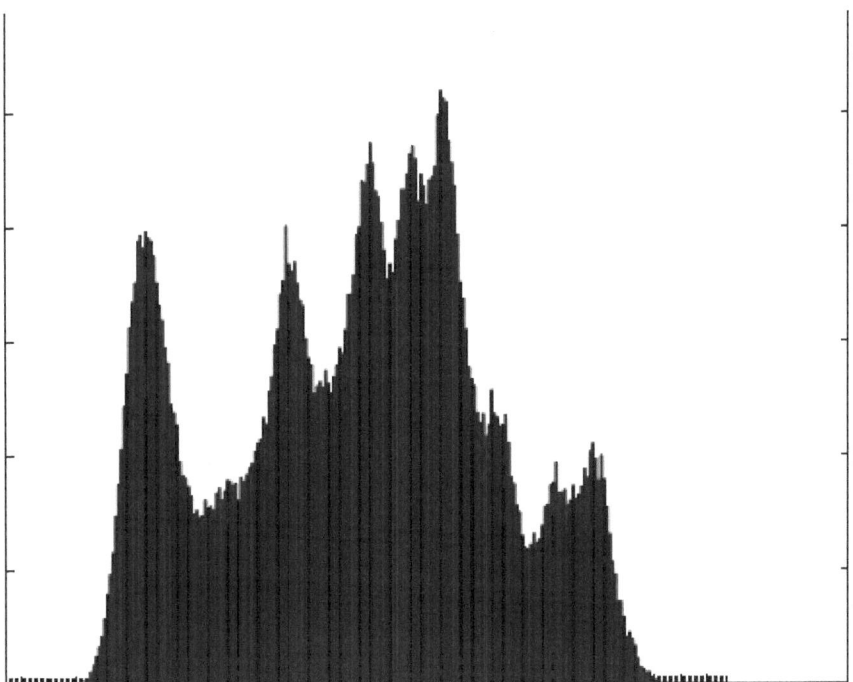

Figure 10 Histogram showing contrast of the Lena image

Many embedded-vision applications require the detection of edges. While edges are easy for the human eye to detect, they can require significant processing in the embedded-vision world. First we must consider the three general types of edges found in images:

- Step – Change in intensity over one or a small number of pixels
- Ramp – Change in intensity over a number of pixels e.g. a gradual increase of pixel value
- Roof – A brief change in intensity before returning to the original intensity

Typically, the resultant images from edge detection are binary in nature (e.g. white or black). There are of course a number of different algorithms to detect edges in our images. Three of the most common are:

- [Sobel](#) – Uses two 3x3 kernels—one for edges in the X direction another for edges in the Y direction—from which the gradient and angle can be determined. This is probably the one of the simplest edge-detection approaches as it is gradient based.
- [Canny](#) – A multi-stage process that uses Gaussian filtering to remove noise. Stages include edge-detection operators like the Sobel operator, non-maximal suppression, thresholding, and hysteresis.
- [LoG – Laplacian of Gaussian](#), applies a Laplacian filter to the results of a Gaussian filter.

Figure 11 Edge-Detection Algorithms Clockwise: Original, Laplacian of Gaussian, Canny, and Sobel

Image Processing With Xilinx Devices

Both Canny and LoG operations are often called advanced edge-detection algorithms because they calculate:

- First derivative – Used to determine the edge
- Second derivative – Used to determine the direction of the edge (e.g. black to white)

Having introduced some even embedded-vision algorithms, we are now going to start looking at how we can use HLS (high-level synthesis) to implement these algorithms within our embedded-vision application.

BUILDING BLOCKS OF AN EMBEDDED VISION SYSTEM

Embedded vision systems are used for a number of applications from simple monitoring systems such as surveillance cameras, to more advanced applications like Advanced Driver Assistance Systems (ADAS) found in today's newest automobiles and Machine Vision used in advanced manufacturing facilities and factories. Regardless of the application there are a number of common elements to an embedded vision system, at a high level these can be grouped as follows:

- Device Interfacing – provides the interface to the selected imaging device. Depending upon the type of device selected this will provide the required clocking, biases and configuration data required. This will also receive the image data from the device decoding it if necessary and formatting it for further processing by the image processing chain.

- Image Processing Chain – receives the image data from the device interface and performs operations such as color filter array interpolation and color space conversion, e.g. color to grayscale. It is also within the image processing chain that we apply many algorithms to the received images. This may be simple algorithms such as noise reduction or edge enhancement to much more complex algorithms like object recognition or optical flow. It is common to call the algorithm implementation the upstream section of the image processing chain. The complexity of the upstream image processing chain implemented depends, of course, upon the application being implemented. While the output formatting which converts the processed image data into the correct format to be output to either a display or over a communication interface is referred to as the downstream section.

- System Supervision and Control – This is separate to the device interfacing and image processing chain, and provides system supervision and control in two areas. The first area is within the device where it provides
    - Configuration of the image processing chain
    - Providing analytics on the image

Image Processing With Xilinx Devices

- Updating the image processing chain as required during algorithm execution

The second area is control and management of the wider embedded vision system where it provides

- Power management and sequencing of image device power rails
- Performing self-test and other system management functions
- Network enabled or point-to-point communication
- Configuration of the image device over an I²C or SPI link prior to the first imaging operations

Some applications may also allow the system supervision to be able to access a frame store and execute algorithms on the frames within. In this case the system supervision is capable of becoming part of the image processing chain.

These areas require different implementation techniques resulting from the challenges inherent with each stage. Both the device interface and the image processing chain require the ability to process high data bandwidths both internally to implement the image processing chain and externally to transmit the image data from the system. While the system supervision and control requires the ability to processes and respond to commands received over the communication interface and provide support for external communications. If the system supervision is to form part of the image processing chain as well, then a high performance processor is required.

As such traditionally embedded vision systems are implemented using a FPGA / processor combination or increasingly a system on a chip (SoC) device which combines a high performance processor with a FPGA. Let's take a look at the different challenges within each of the three high level areas before we demonstrate how these areas come together in a demonstration.

Device Interface

The sensor interface is determined by the device selected for your application, most embedded vision applications use CMOS Imaging Sensors (CIS). Typically, these sensors use either a CMOS parallel output bus with flags to indicate the line and frame valid sequence or use serialized communications at higher speeds enabling simpler system interfacing but introduce a slightly more complex FPGA implementation. These serialized data streams transmit images over a fewer number of lanes compared to a parallel bus, as they operate at much faster data rates thus enabling the imager to support faster frame rates compared to a parallel interface. To enable synchronization, it is common to have data channels which contain the image and other data words coupled with a synchronization channel which contains code words defining the content on the data channel. Along with data and synchronization lanes there is also a clock lane as the interface is source synchronous. These high speed serialized lanes are normally implemented as LVDS or on Reduced Swing LVDS to reduce system noise and power.

Regardless of the output image format it is also common with CIS devices for them to require configuration by the embedded vision system before any image can be obtained. This is a result of the versatility of CIS devices which provide significant on chip processing which needs to be configured with the correct settings before an image can be output. These interfaces are not as bandwidth critical as the image transfer therefore, interfaces standards such as I^{2}C or SPI are commonly used.

As the image data bandwidth required is high, it is common to implement this interface in a FPGA, this also allows for easier integration with the image processing chain. The CIS device configuration interface typically uses either I^{2}C or SPI and they may be implemented by either the FPGA or by the system supervision and control processor if that supports the interface required.

Image Processing Chain

The image processing chain consists of both the upstream and downstream elements and interfaces with the pixel data output by the device interface. However, the pixels received may not be in the format which can be used to correctly display an image. We may need to perform image correction; this is especially the case if a color imager is used.

To maintain throughput at the data rates required, image processing chains are often implemented in FPGAs exploiting their parallel nature.

Image Processing With Xilinx Devices

This allows an image processing pipeline to be generated allowing each step of the processing chain to be implemented concurrently, allowing a higher frame rate to be achieved. However, for some applications we must consider the latency as well, this is especially the case for systems like Advanced Drivers Assistance Systems (ADAS). To establish an image processing chain efficiently, we need to base the image processing cores around a common interconnect protocol allowing for the processing IP to be interconnected with ease. This provides two benefits, a reusable library of components and allows the establishment of the pipeline with ease as each IP core is designed to receive and transmit data according to a defined standard. There are a number of commonly used protocols with one of the most common being AXI due to its flexible nature supporting both memory mapped and streamed interfaces.

Typical processing stages within the image processing chain are:

- Color Filter Array – Generation of each pixels color which results from the Bayer pattern on the CIS device
- Color Space Conversion – Conversion from RGB to YUV used in many image processing algorithms and output schemes
- Chroma Resampling – Conversion of the YUV pixels to a more efficient pixel encoding
- Apply image correction algorithms such as color or gamma correction, or perform image enhancement or noise reduction
- On the downstream side we can configure the video output timing and then convert back to native parallel output video format prior to it being output to the required driver

Some systems also use external memory, typically DDR-based as a frame store, within a SoC this is often available to the processor side of the SoC as well. This ability to share the memory space allows the system supervision side to transfer data over networks like Gigabit Ethernet or USB or become an extension to the image processing chain.

System Supervision

This is traditionally implemented in the processor, providing the ability to process commands to configure the image processing chain to the demands of the application. To enable commands to be received and processed the system supervision must be capable of supporting a number of different communications interfaces from simple RS232,

Giabit Ethernet, USB, PCIe and more specialized interfaces like CAN (used in automotive applications).

Provided the architecture of the embedded vision system supports it we can use the processor to generate image overlay information which maybe superimposed on the output image. With access to the image data we can also use the processor to perform additional processing upon the image or to gather statistics such as histograms of pixel value distribution. Allowing this to become part of the image processing chain and enabling the developer to exploit a range of open source image processing libraries like OpenCV, OpenCL and OpenVX.

Example Using EVK

Having explained the basic elements of an embedded vision system it is often good to demonstrate these concepts and show how they are combined to create a working system. This example will show how we can create an embedded vision system using the Avnet MicroZed Embedded Vision Kit (EVK).

This kit is uses the On Semi Python 1300C imaging device and the Xilinx Zynq 7020. The Python 1300C is a color device with 1280 pixels by 1024 lines and the device is configured over a SPI interface. The variant of sensor provided uses a serialized output to enable high frame rates, while the EVK supports image output over the HDMI interface to a display.

The Zynq 7020 provides a great platform for the embedded vision system's implementation as we can use the FPGA fabric, called the PL for Programmable Logic to implement the device interface and image processing chain. While using the dual ARM A9 cores called the PS for Processing System for system supervision and image processing chain extension if we so desire.

To develop the application, we will be using two SoC development tools. The first is Xilinx Vivado and the second is Xilinx SDK. Within Vivado we will be implementing the device interface, the image processing chain, configure the PS within the Zynq and establish the PS to PL memory mapped communications to perform the following operations:

Image Processing With Xilinx Devices

- Configure the IP within the image processing chain with the parameters required for the image sizes and frame rates and required operations. To do this we will use the general purpose AXI interconnect between the PS and PL with the PS as the master. Using this interface we can transfer 1,200 Mbps between the PS and the PL.
- Place the processor's DDR memory within the image processing chain such that if necessary the processor can access it. To do this we will be using the high performance AXI interconnect between the PL and the PS where the PL is the master, using this interface we can transfer 2,400 Mbps between the PL and the PS DDR Memory.

This demonstration will output the image to a display using HDMI. Rather helpfully, Avnet, the makers of the EVK, supply a device interface IP module for the Python 1300C and a HDMI output IP module to interface with the output HDMI device on the EVK. We will be using these with our example. Within Vivado we can add in these IP modules into the Vivado IP Catalog using the IPXact format.

The image processing chain will interface to the On Semi device interface and perform the following stages, with the exception of the Python 1300C and HDMI IP cores all of the IP cores used are contained within the standard Xilinx Image Processing IP library in Vivado (the names of the actual cores are in italics below):

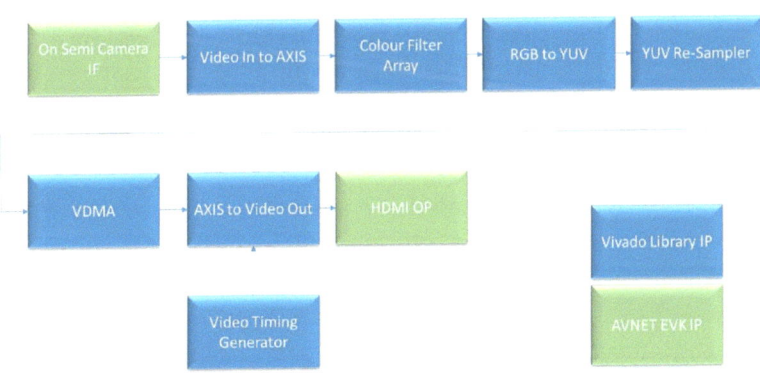

Figure 12 Image Processing Chain

- Conversion of the parallel video and horizontal and vertical syncs from the Python Interface IP into an AXIS stream such that

we are able to interface it with the subsequent image processing IP cores. *Video in to AXIS*
- Color Filter Array enables each output pixel to be given a RGB value by using the information from the Bayer pattern covering the input pixels (which only represent either R, G or B). *Color Filter Array interpolation*
- RGB to YUV color space conversion, convert into a different color space as this is the output format preferred by the HDMI driver. *RGB to YCRCB Color-Space Convertor*
- Rescale the YUV from 4:4:4 to 4:2:2 format. *Chroma Resampler*
- Configure a Video DMA to transfer the Images frames to the PS DDR. *AXI VDMA*
- Configure the same Video DMA to read images frames from the PS DDR. *AXI VDMA*
- Convert the AXI Stream back to Parallel format. *AXIS to Video Out*
- Provide a timing reference generator for the output video timing. *Video Timing Controller*

In addition to ensure the system functions correctly we also require two *AXI Interconnects*. One for the high performance AXI interconnect, and another for the general purpose AXI Interconnect, along with the necessary reset blocks for each of the clock domains.

Image processing applications require a number of clock domains, for most of those required we can use the PL fabric clocks provided by the PS within the Zynq. For this application we require the following clocks:

- 108 MHz – This is the pixel clock rate for an image being output at 1280 by 1024 at 60 Hz.
- 75 MHz – Used for the memory mapped AXI and AXI lite interfaces.
- 150 MHz – Used for the image processing chain also known as the AXI Streaming clock. The AXI Stream clock must be at least equal to the pixel rate. However, we must consider the throughput of all IP cores within the processing chain. While most of the modules are capable of processing one pixel per clock, it is wise to have some margin and reduce the buffering required.

Image Processing With Xilinx Devices

- 200 MHz – Supplied to the Python 1300C CIS device as reference.

To achieve the pixel clock we need to use a clock wizard to generate the 108 MHz pixel clock as this needs to be set very accurately. While the 75 MHz and 150 MHz have some tolerance when set by the PL fabric clocks, the 200 MHz clock also requires accuracy. Unlike the 108 MHz clock this can be generated accurately by the PS fabric clocks.

The clocking structure of the IP modules used is demonstrated in the table below.

IP Module	AXI Lite / MM clock	AXI Streaming Clock	Pixel Clock	Additional Clocks
On Semi Python 1300C	NA	NA	108 MHz	200 MHz
On Semi SPI	75 Mhz	NA	NA	NA
Video in to AXIS	NA	150 MHz	108 MHz	NA
Color Filter Array interpolation	75 MHz	150 MHz	NA	NA
RGB to YCRCB Color-Space Convertor	NA	150 MHz	NA	NA
Chroma Resampler	NA	150 MHz	NA	NA
AXI VDMA	75 MHz	150 MHz	NA	NA
AXIS to Video Out	NA	150 MHz	108 MHz	NA
Video Timing Controller	NA	NA	108 MHz	NA

| HDMI OP | NA | NA | 108 MHz | NA |

Table 2 Image Processing chain clocking structure

The complete implementation in the Vivado block diagram editor can be seen in the two figures below which identify the upstream and downstream image processing chains. You can also see the interconnections to and from the ARM core processors using the general purpose and high performance AXI interconnects.

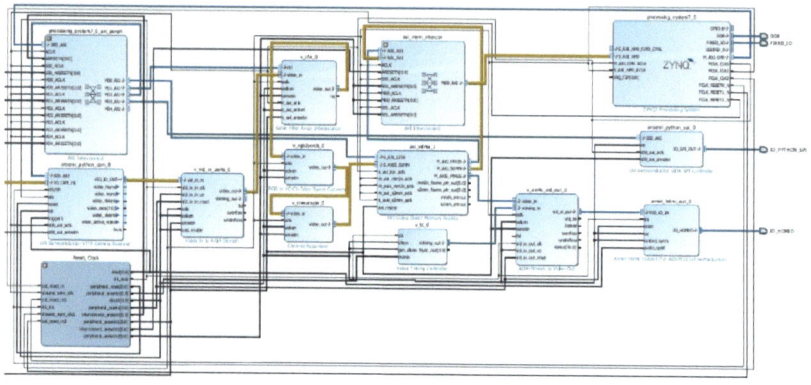

Figure 13 EVK Example Design – The highlighted path is upstream image processing chain

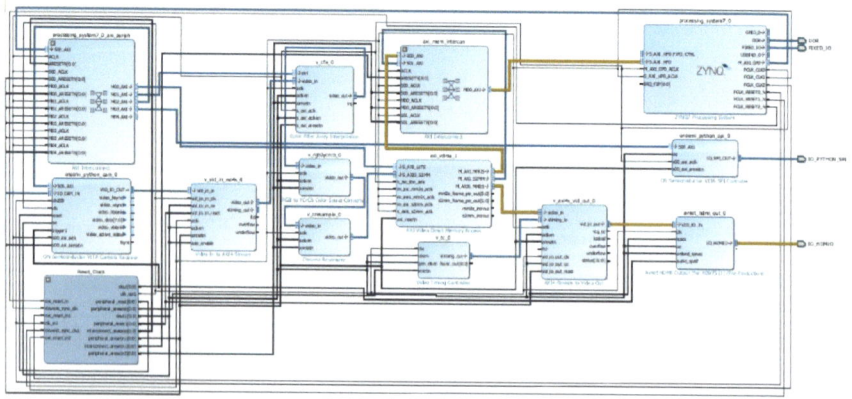

Figure 14 EVK Example Design – The highlighted path is the downstream image processing chain

Once we have verified the design and assigned addresses to the memory mapped peripherals which use the high performance and

Image Processing With Xilinx Devices

general purpose AXI interconnects (this can be done automatically), we can build the hardware in Vivado 2015.4 and export it to the SDK 2015.4 software development environment where we need to write some simple software to get the system up-and-running.

Within SDK we need to configure not only the design within the Zynq but also some elements on the EVK, prior to using it. Remember in this example the PS is acting as the System supervisor and control function, and hence must also configure the entire embedded vision system and not just the Vivado design within the Zynq.

We need to develop software which configures

- The Python 1300C camera using its SPI interface
- The AXI Python 1300C interface module
- The AXI VDMA, to read from and write frames to DDR memory
- The AXI Color Filter Array
- The HDMI Output device for the AD7511 this uses I^2C for configuration
- The I^2C Mux and its attached peripherals of interest
- The I^2C IO expander controlling power rails for the Python 1300C device

The EVK uses the Zynq PS I^2C controller to configure the HDMI output device and enable the power supplies to the Python device. Avnet supplies APIs which we can use to control the I^2C and configure the following:

- ADV7511 – API for the HDMI output
- CAT9554 – API for the I^2C I/O expander on the camera module
- TCA9548 – API for the I^2C mux on the EVCC
- PCA9534 – API for the I^2C IO expander on the EVCC
- OnSemi_Python_SW – API for the Python 1300C
- XAXIVDMA_EXT – API for configuring the VDMA
- XIICPS_EXT – API for driving the external I^2C

All we need to do is use these coupled with the Xilinx software APIs for the IP within the image processing chain and we can therefore quickly create the software executable. To create the software application, we need to import the hardware design into SDK and create a board support package (BSP) for this hardware. This BSP will contain all of the Xilinx APIs needed when coupled with the Avnet APIs to drive the hardware in the Image processing chain and the Zynq.

The software itself is required to perform the following steps

- Initialize all of the AXI peripherals

```
piicps_config = XIicPs_LookupConfig(XPAR_XIICPS_0_DEVICE_ID);
XIicPs_CfgInitialize(pdemo->piicps0, piicps_config, piicps_config->BaseAddress);
XIicPs_Reset(pdemo->piicps0);
XIicPs_SetSClk(pdemo->piicps0, 100000);

paxivdma_config = XAxiVdma_LookupConfig(XPAR_AXIVDMA_0_DEVICE_ID);
XAxiVdma_CfgInitialize(pdemo->paxivdma0, paxivdma_config, paxivdma_config->BaseAddress);

pcfa_config = XCfa_LookupConfig(XPAR_V_CFA_0_DEVICE_ID);
XCfa_CfgInitialize(pdemo->pcfa, pcfa_config, pcfa_config->BaseAddress);

onsemi_python_init(pdemo->pPython_receiver, "PYTHON-1300-C",
        XPAR_ONSEMI_PYTHON_SPI_0_S00_AXI_BASEADDR, XPAR_ONSEMI_PYTHON_CAM_0_S00_AXI_BASEADDR);
onsemi_python_spi_config(pdemo->pPython_receiver,4);
pdemo->pPython_receiver->uManualTap = 25; // IDELAY setting (0-31)
```

- Power up the image sensor rails

```
// Make sure all disable first
cat9554_vddpix_off(pdemo->piicps0);
usleep(10);
cat9554_vdd33_off(pdemo->piicps0);
usleep(10);
cat9554_vdd18_off(pdemo->piicps0);
usleep(1000);

// Turn them on one by one
cat9554_vdd18_en(pdemo->piicps0);
usleep(10);
cat9554_vdd33_en(pdemo->piicps0);
usleep(10);
cat9554_vddpix_en(pdemo->piicps0);
usleep(10);
```

- Configure the On Semi Python 1300C, the Color Filter Array and the VDMA

```
onsemi_python_sensor_initialize(pdemo->pPython_receiver, SENSOR_INIT_ENABLE, 0);
onsemi_python_sensor_initialize(pdemo->pPython_receiver, SENSOR_INIT_STREAMON, 0);
onsemi_python_sensor_cds(pdemo->pPython_receiver, 0);
onsemi_python_cam_reg_write(pdemo->pPython_receiver,(Xuint32) ONSEMI_PYTHON_CAM_SYNCGEN_HTIMING1_REG, 0x00300500);

printf("CFA Initialization\r\n");
XCfa_Reset(pdemo->pcfa);
XCfa_SetBayerPhase(pdemo->pcfa, XCFA_RGRG_COMBINATION);
XCfa_RegUpdateEnable(pdemo->pcfa);
XCfa_Enable(pdemo->pcfa);

printf("VDMA Initialization\r\n");
XAxiVdma_Reset(pdemo->paxivdma0, XAXIVDMA_WRITE);
XAxiVdma_Reset(pdemo->paxivdma0, XAXIVDMA_READ);
WriteSetup(pdemo->paxivdma0, 0x10000000, 0, 1, 1, 0, 0, 1280, 1024, 2048, 2048);
ReadSetup(pdemo->paxivdma0, 0x10000000, 0, 1, 1, 0, 0, 1280, 1024, 2048, 2048);
StartTransfer(pdemo->paxivdma0);
```

Image Processing With Xilinx Devices

Having completed these steps when the software is run on the EVK you will see an image being output on your chosen HDMI monitor as below.

Figure 15 Frame grab of a residential scene captured using the demo

USING A ZYNQ WITHIN YOUR EMBEDDED VISION SYSTEM

Embedded Vision Systems require the ability to implement a high performance image processing chain along with implementing control and communication functions. Traditionally embedded systems have implemented this using separate a FPGA and Processor, such an approach leads to a more complicated design and increases the Size, Weight and Power – Cost (SWAP-C) of the system. As the traditional design is more complicated it also requires more Non-Recurring Engineering development and takes longer to bring to market.

As embedded vision architects we can address the issues off SWAP-C, NRE and time to market by using the Zynq All Programmable System On Chip. This device combines dual ARM A9 cores with seven series FPGA logic allowing us to implement our embedded vision system within a single device.

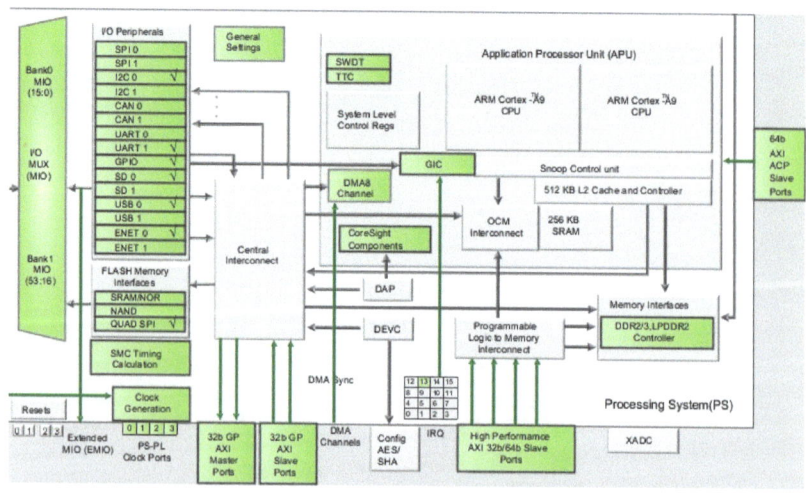

Figure 16 Zynq Architecture

The Zynq is the ideal platform for any embedded system it provides a number of industry standard IO peripherals which enable data to be moved efficiently on and off chip, while also providing support for a number of volatile (DDR2/3/LP) and non-volatile memories.

Image Processing With Xilinx Devices

For embedded vision systems there are four main aspects which make the Zynq ideal beside the combination of processor and FPGA fabric.

- Internal Data Movement Architecture
- Support for multiple system interfaces
- IP libraries for the image processing chain
- Software Development Environment

Let's take a look at each of these areas in turn and explore how they can be used to create an embedded vision system.

Internal Data Movement Architecture

Zynq Architecture is based around the ability to move data at high bandwidths not only between the processor system (PS) and the programmable logic (PL) sides of the device but also within the PS system and PL fabric. To achieve this performance, the embedded vision system designer is provided with an internal infrastructure based upon the Advanced eXtensible Interface (AXI) infrastructure.

The Zynq architecture as shown in figure one is interconnect rich, however to gain the maximum performance we need to utilise the correct interconnect structures. Focusing initially on the PS/ PL interface, we have the choice of the following interconnects

- Four General Purpose AXI (GP AXI) interconnects with two masters in each direction, this is a 32-bit interface and is connected to the central interconnect within the PS allowing providing a wide range of data transfer sources and destinations.
- Four High Performance AXI (HP AXI) interconnects, these are designed to move data efficiently from the PL to the PS particularly On Chip Memory (OCM) and DDR (PS DDR) memory. As such the PL is the master importantly these support 64 bit transfers.
- One Accelerator Coherence Port AXI (ACP AXI) this provides a 64-bit cache coherent interconnects from the PL to the SCU.

These three different methods of communicating between PL and PS provide the ability to move 16,800 MBps of data between the PL and PS. This provides a significant advantage for embedded vision system development which are both throughput and bandwidth intensive.

Interface	IF Clock	Read BW	Write BW	No Ports	Total BW
AXI GPIO	150 MHz	600 MBps	600 MBps	4	4800 MBps
AXI HP	150 MHz	1200 MBps	1200 MBps	4	9600 MBps
AXI ACP	150 MHz	1200 MBps	1200 MBps	1	2400 MBps

Table 3 PS / PL interfaces and bandwidths

Efficient data transfer between the PS and PL requires the use of Direct Memory Access (DMA) within the Zynq we can use either the PS DMA or AXI DMA to transfer the data efficiently depending upon which is the master PS or PL. When we consider embedded vision applications DMA coupled with the HP AXI ports enable the image to be transferred to and from the PS DDR memory enabling it to act as a frame buffer. Of course once the image frames are within the PS DDR they can also be accessed and post processed by algorithms running in the PS and either output using system interfaces or re written to the frame buffers for output as shown in figure two.

Within the PL Fabric the use of AXI between design modules is important as it allows the provision of both simple configuration and control of modules from the PS allowing real time configuration as required and allowing the interfacing of high performance modules with ease. The use of AXI also allows the leveraging of Xilinx and Alliance partner IP which allows for a reduction in development time and allows the developers the ability to create a library of their own IP which can be reused across several programmes. Within the PL we can implement three flavours of AXI

Image Processing With Xilinx Devices

- AXI Lite – Used for low performance memory mapped applications for example configuration and monitoring of IP modules.
- AXI – Used for high performance memory mapped applications, this allows large bursts of data to maximise throughput – a good example of this is the AXI Video DMA which converts to and from AXI Streaming to AXI Memory Mapped.
- AXI Streaming – Unidirectional interfaces which are not memory mapped, with unlimited burst length, AXI streaming interfaces are common within the image processing chain.

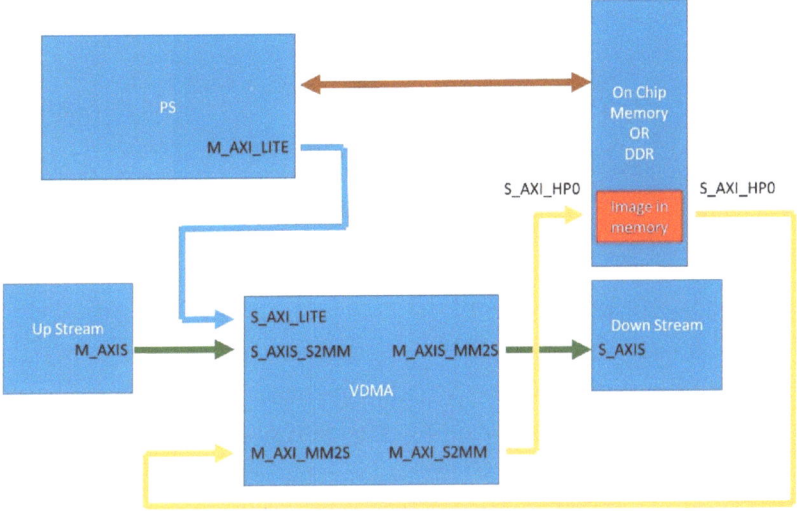

Figure 17 Using AXI streaming, AXI and DMA to move images between PS and PL (Green = AXI Streaming, Yellow = Memory Mapped AXI, Blue = AXI Lite, Orange = PS / DDR link)

System Interfaces

The ability to move data efficiently around the Zynq to create the image processing pipeline is just one challenge, embedded vision systems are required to output the image to a display system, other networked processing elements or a storage medium. The Zynq enables the embedded system designer a wide range of industry standard communication peripherals. These peripherals can be split between high performance peripherals and standard peripherals which can be used to control and output images and data from the system. The high performance peripherals include

- 2 x Gigabit Ethernet within the PS
- 2 x USB 2.0 within the PS
- 2 x SD / SDIO within the PS
- GTX or GTP transceivers at up to 12.5 GBps (GTX) or 6.25 GBps (GTP) located within the PL
- PCIe end point or root complex, located within the PL

These high performance IO peripherals located within the PS also have dedicated DMA controllers within the peripheral allowing for more efficient data movement within the PS and a reduced processor loading.

While the standard peripheral support includes

- 2 x UART – These can be used to communicate externally using RS232, R422 etc.
- 2 x I2C – Used for on PCB communication with peripheral components and often the image sensor
- 2 x SPI – Used for on PCB communication with peripheral components and often the image sensor
- 2 x CAN – Used for communication externally, this is a very popular interface for automotive and industrial applications

Of course the PS also provide general purpose IO should the need arise, while the PL IO comes in two flavours High Range (HR) and High Performance (HP) which support a number of IO standards and including LVDS which allows for implementations of IO standards like Camera Link if point to point image output is required.

Provision of the GTP and GTX transceivers within the PL also allow for the inclusion of 10 Gigabit Ethernet using external PHY, this allows for both electrical or optical implementation to be implemented. This interface is used in many Storage Area Networks (SAN) and shows the versatility of the GTP and GTX ports for providing high bandwidth flexible interfacing.

While PCIe capability also provided within the PL provides for the ability

Image Processing With Xilinx Devices

to interface within a more traditional back plane architecture if that is required by the embedded vision system architecture.

These interfaces allow us to implement embedded vision systems that are capable of outputting the images at the frame rates required over a number of different standards removing what has previously been a bottleneck.

IP Libraries

It is within the PL side of the device that we will implement most of the image processing chain, using the PS DDR as a frame store we can split the image processing chain to be upstream or downstream

- Upstream refers to the image processing chain prior to the DMA into the PS DDR
- Downstream refers to the image processing chain after the DMA from the PS DDR to the output.

Implementing the image processing chain can require complex algorithms be implemented but it also requires the image be first received and formatted as desired and that it can be output in the desired format. To aid the development of the image processing chain the Xilinx IP library provides a number of modules from both Xilinx and Alliance partners which can be customised for most applications.

Figure 18 Image Processing IP cores available

These IP modules are based around the AXI streaming protocol to transfer image data between blocks and often AXI lite / AXI to configure

41

the core from the PS for particular operations. To enable conversion from the native parallel video to AXI Streaming and back again prior to output there are IP modules provided to do this. From the IP blocks provided we can implement an image processing chain with great ease performing most of the functions required, we can for example create an image processing chain which implements

- Color Filter Array Interpolation – reconstruction of the pixel colour on Imagers using a Bayer pattern
- Color Space Conversion – convert from RGB to YUV colour Space as required for many output stands and processing functions.
- Chroma Re Sampling – Re sample the YUV color space to a more efficient pixel encoding
- Apply image correction algorithms such as color or gamma correction, or perform image enhancement or noise reduction.
- AXI Video DMA – Convert the AXI streaming data to a AXI memory map and store it within frame buffers in the PS DDR. AXI VDMA can be either unidirectional or bi-directional providing read and write operations to the PS DDR frame buffers depending upon our architectural needs.
- On the downstream side we can configure the video output timing and then convert back to native parallel output video format prior to it being output to the required driver.

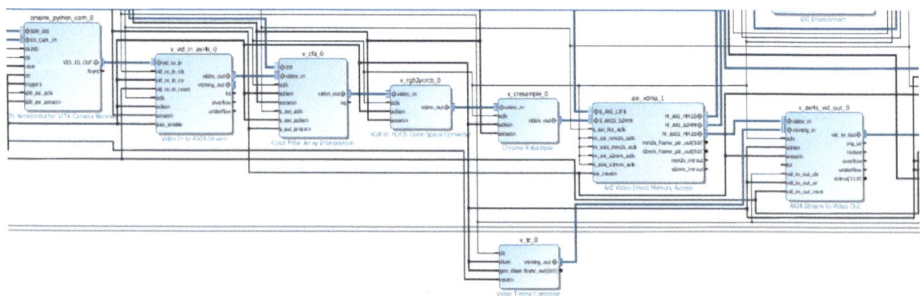

Figure 19 Image Processing Chain Implementation

All we need develop is the interface to our image sensor and any additional image processing that we require and for that there are several tools in the tool box to help generate the processing required utilising the capability of High Level Synthesis (HLS).

SW Development Environment

To obtain the maximum performance from the system we should use the SDSoC development environment, this allows us to develop traditional applications like we would within the SDK environment. However, SDSoC also enables us to leverage the potential of the HLS and the connectivity frame work to design functions within C or C++ and then accelerate the function in the PL of the device. For image processing this becomes very important as we can leverage the power created by image processing libraries like Open CV and Open VX to quickly and efficiently implement algorithms in C/C++ and move them into the image processing chain using the HLS, coupled with AXI VDMA makes for a very powerful image processing system.

The development flow for a typical SDSoC Application is shown below

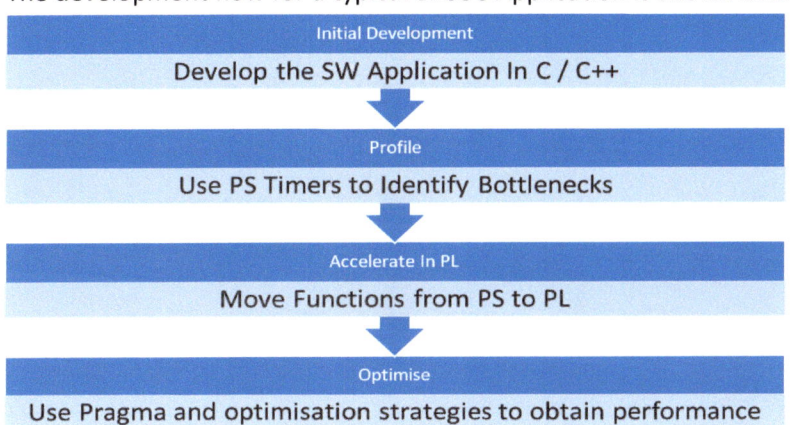

Figure 20 SDSoC Development Flow

At the heart of the SDSoC platform is the connectivity framework, this enables the acceleration of the functions and integration with the overall application. The connectivity frame work uses four stages to implement the solution

- Perform HLS on the function to be accelerated

- Analyse the communication to and from the accelerated function
- Establish a communication method using the available AXI Ports between the PS and PL, to maximise performance DMA structures are preferred
- Generate the necessary software files to replace the accelerated function and handle communicate to and from it.

We can iterate the profile, accelerate and optimise stages of the SDSoC Approach to ensure we obtain the performance we require for the system. Optimisation is performed using pragmas within the source code and enables us to direct HLS to optimise as required.

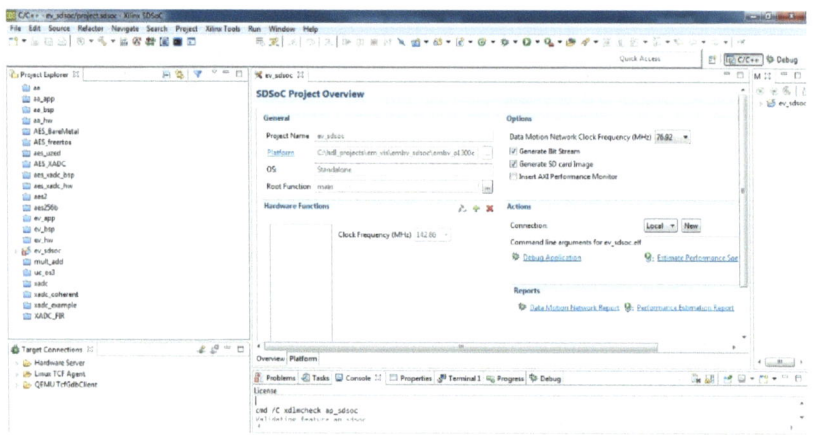

Figure 21 SDSoC – eclipse based development environment

Of course not all embedded vision systems require further processing instead the images within the PS DDR frame maybe transmitted out over the one of the IO peripherals in essence acting as a frame grabber. Here we can leverage the operating systems supported and use Linux or FreeRTOS / bare metal coupled with the Light Weight IP stack to transfer data from the frame stores out over the Gig Ethernet or other supported interface.

USING A FPGA WITHIN YOUR EMBEDDED VISION SYSTEM

Over this blog series, I have written a lot about how we can use the Zynq SoC in our designs. We have looked at a range of different applications and especially at embedded vision. However, some systems use a pure FPGA approach to embedded vision, as opposed to an SoC like the members in the Zynq family, so in this blog we are going to look at how we can get a simple HDMI input-and-output video-processing system using the Artix-7 XC7A200T FPGA on the Nexys Video Artix-7 FPGA Trainer Board. (The Artix-7 A200T is the largest member of the Artix-7 FPGA device family.)

Here's a photo of my Nexys Video Artix-7 FPGA Trainer Board:

Figure 22 Nexys Video Artix-7 FPGA Trainer Board

For those not familiar with it, the Nexys Video Trainer Board is intended for teaching and prototyping video and vision applications. As such, it comes with the following I/O and peripheral interfaces designed to support video reception, processing, and generation/output:

- HDMI Input
- HDMI Output
- Display Port Output
- Ethernet
- UART
- USB Host
- 512 MB of DDR SDRAM
- Line In / Mic In / Headphone Out / Line Out
- FMC

To create a simple image-processing pipeline, we need to implement the following architecture:

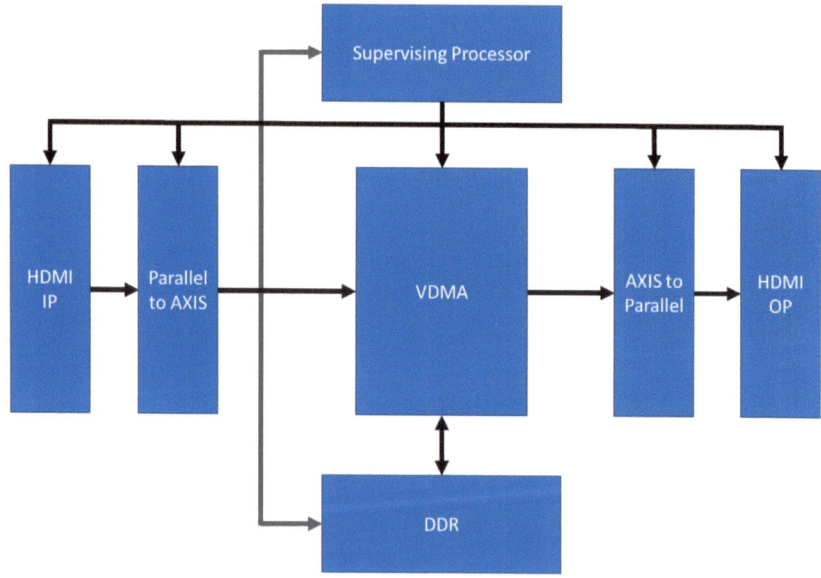

Figure 23 Architecture of the Image processing system in the FPGA

The supervising processor (in this case, a Xilinx MicroBlaze soft-core RISC processor implemented in the Artix-7 FPGA) monitors communications with the user interface and configures the image-processing pipeline as required for the application. In this simple architecture, data received over the HDMI input is converted from its parallel format of Video Data, HSync and VSync into an AXI Streaming (AXIS) format. We want to convert the data into an AXIS format because

Image Processing With Xilinx Devices

the Vivado Design Suite provides several image-processing IP blocks that use this data format. Being able to support AXIS interfaces is also important if we want to create our own image-processing functions using Vivado High Level Synthesis (HLS).

The MicroBlaze processor needs to be able to support the following peripherals:

- AXI UART – Enables communication and control of the system
- AXI Timer – Enables the
- MicroBlaze Debugging Module – Enables the debugging of the MicroBlaze
- MicroBlaze Local Memory – Connected to DLMB and ILMB (Data & Instruction Local Memory Bus)

We'll use the memory interface generator to create a DDR interface to the board's SDRAM. This interface and the SDRAM creates a common frame store accessible to both the image-processing pipeline and the supervising processor using an AXI interconnect.

Creating a simple image-processing pipeline requires the use of the following IP blocks:

- DVI2RGB – HDMI input IP provided by Digilent
- RGB2DVI – HDMI output IP provided by Digilent
- Video In to AXI4-Stream – Converts a parallel-video input to AXI Streaming protocol (Vivado IP)
- AXI4-Stream to Video Out – Converts an AXI-Stream-to-Parallel-video output (Vivado IP)
- Video Timing Controller Input – Detects the incoming video parameters (Vivado IP)
- Video Timing Controller Output – Generates the output video timing parameters (Vivado IP)
- Video Direct Memory Access – Enables images to be written to and from the DDR SDRAM

The core of this video-processing chain is the VDMA, which we use to move the image into the DDR memory.

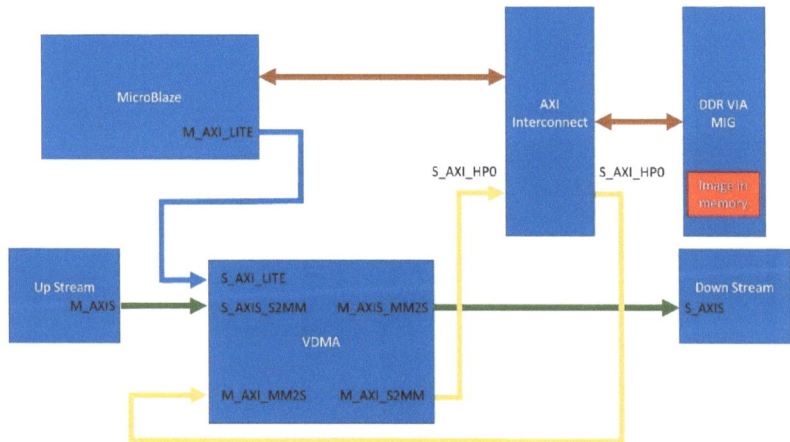

Figure 24 Data Flow

The diagram above demonstrates how the IP block converts from streamed data to memory-mapped data for the read and write channels. Both VDMA channels provide the ability to convert between streaming and memory-mapped data as required. The write channel supports Stream-to-Memory-Mapped conversion while the read channel provides Memory-Mapped-to-Stream conversion.

When all this is put together in Vivado to create the initial base system, we get the architecture below, which is provided by the Nexys Video HDMI example.

Image Processing With Xilinx Devices

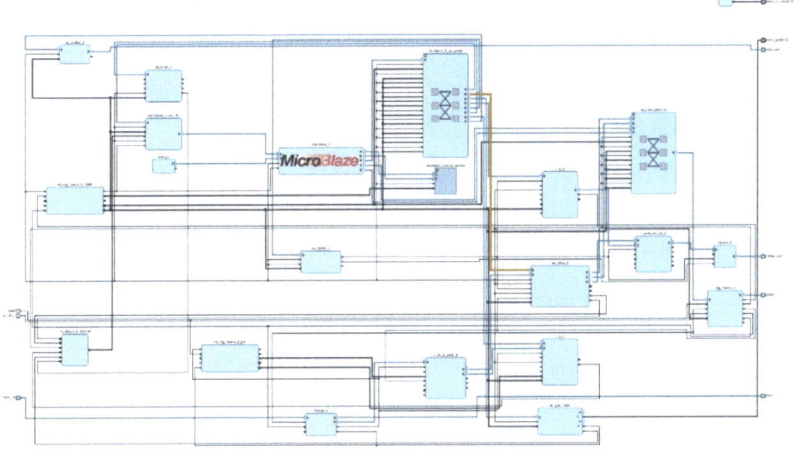

Figure 25 Finalised design

All that is required now is to look at the software required to configure the image-processing pipeline.

With the MicroBlaze system up and running on the Nexys board we need some software to generate a video output. In this example, we are going to use the MicroBlaze to generate test patterns, to do this it will write data into the DDR such that the VDMA can read this out and output it over HDMI.

Within the SW the first thing we will need is to define the frames which are going to be stored in VDMA and output to do this we will define three frames within DDR Each frame will be defined as a two-dimensional array

u8 frameBuf[DISPLAY_NUM_FRAMES][DEMO_MAX_FRAME];

Where the DISPLAY_NUM_FRAME is set to 3 and DEMO_MAX_FRAME is set to 1920 * 1080 *3 this is to take account of the max frame resolution. The final multiplication by 3 is required as each pixel red, Green and Blue is represented by 8 bits.

To access these frames, we can use an array of pointers to the each of the three frame buffers, this will ease our interaction with the frames.

With the frames defined the next stage it is to initialise and configure the peripherals within the design these are:-

- VDMA – Uses DMA to move data from the DDR to the output video chain
- Dynamic Clocking IP – This outputs the pixel clock frequency and multiples of this for the HDMI output.
- Video Timing Controller 0 – This defines the output display timing depending upon resolution
- Video Timing Controller 1 – This determines the video timing on the input received, in this demo this is used to grab input frames from a source

To ensure the VDMA functions correctly we need to define the stride this is the separation between each line within the DDR memory as such for this application it is 3 * 1920 which is the maximum length of a line.

When it comes to the application we will be able to set different display resolution from 640x480 to 1920x1080.

```
************************************************************
*              Nexys Video HDMI Demo                        *
************************************************************
*Current Resolution:                       640x480@60Hz*
*Pixel Clock Freq. (MHz):                         25.000*
************************************************************

1 - 640x480@60Hz
2 - 800x600@60Hz
3 - 1280x720@60Hz
4 - 1280x1024@60Hz
5 - 1920x1080@60Hz
q - Quit (don't change resolution)

Select a new resolution:
```

Figure 26 test application running

No matter what the resolution selected we will be able to draw test

Image Processing With Xilinx Devices

patterns on the screen using software functions which are written into the DDR. When we change functions, we need to reconfigure the VDMA, Video Timing Generator 0 and the dynamic clocking module.

With these configured the next stage is to generate video output, with this example there are many functions within the main application which generate, capture and display video, these are:-

1. Bar Test Pattern – Generates several colour bars across the screen
2. Blended Test Pattern – A blended test pattern across the screen
3. Streaming from the HDMI input to the output
4. Grab an input frame and inverting the colours
5. Grab an input frame and scale to the current display resolution

Within each of these functions we pass the pointer to the frame currently being output such that we can modify the pixel values in memory. This can be done simply as shown in the code snippet below which sets the Red, Blue and Green pixels, where each one is unsinged 8 bits.

```
frame[iPixelAddr]     = wRed;
frame[iPixelAddr + 1] = wBlue;
frame[iPixelAddr + 2] = wGreen;
```

When we run the application, we can choose which of the functions we want to exercise using the menu output over the UART terminal.

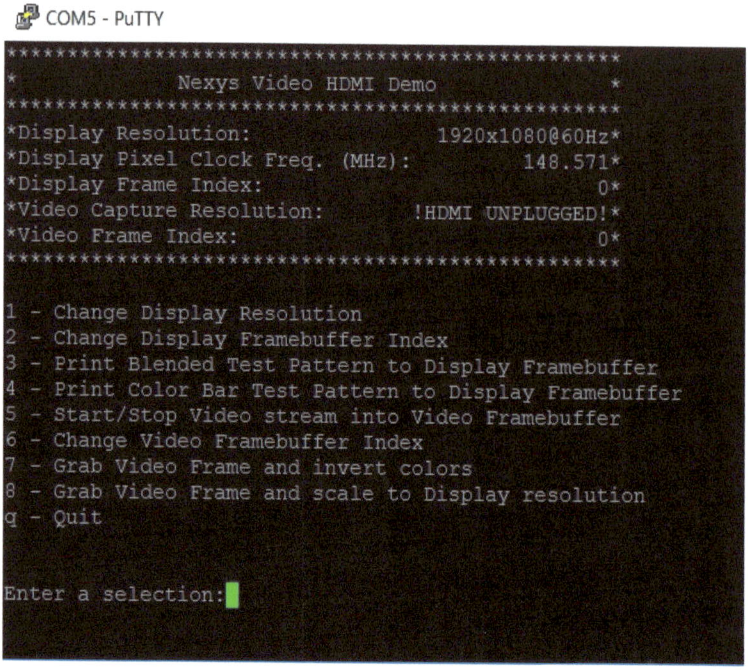

Figure 27 Test application control over UART

Setting the program to output the color bars and the blended test gave the outputs below on my display.

Image Processing With Xilinx Devices

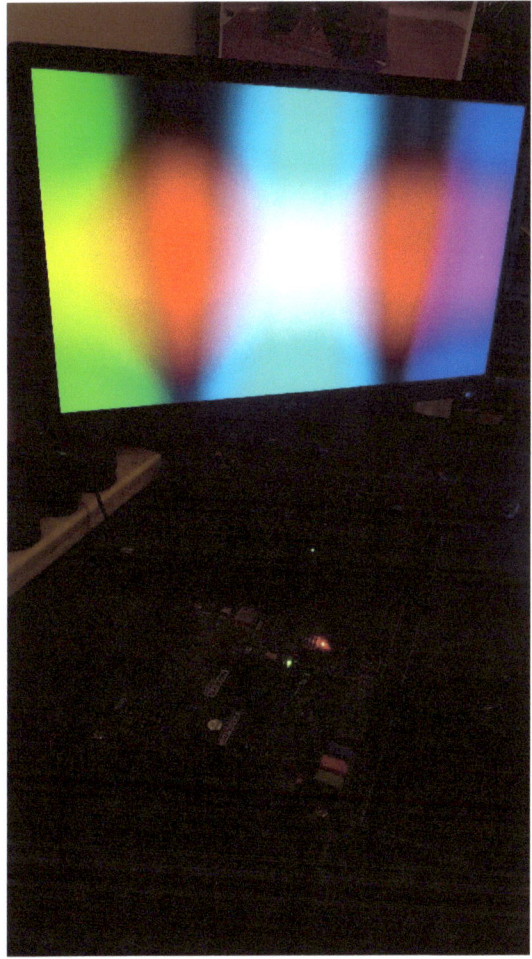

Figure 28 Blended Test Pattern Output

Figure 29 Color bars test pattern

Now that we know how we can write information to DDR memory and see it appear on our display if we so desire.

INTERFACING: CAMERA LINK

I thought it would be a good idea to examine how we can interface actual image sensors to the Zynq SoC so that we can obtain an image that we can process.

At a high level, we can break down the interface into one of two different categories:

- Camera Interface – We wish to interface to the video output port of an existing camera. In this case, the output video may use a protocol like Camera Link, USB or GigE Vision.

- Image Sensor Interface – We wish to interface directly to the image sensor. In this case, the interface may be a parallel bus or a high-speed LVDS bus depending upon the sensor chosen.

This is where the flexibility of the Zynq SoC really comes in handy. The ability to use the embedded peripheral cores in the Zynq SoC's PS and the programmable I/O and logic in the Zynq SoC's PL allow you to interface your design to any camera or any sensor and to create a tightly integrated system. The programmable nature of these interfaces means that you can use the Zynq SoC to create a vision platform for many varied camera and image-processing designs, and several commercial camera vendors have done exactly that.

If we are interfacing to a camera with a USB or GigE Vision video interface, we can use the I/O peripherals in the Zynq SoC's PS. Images captured over these interfaces can then be routed via the central interconnect of the PS directly into attached DDR memory. Once the image is stored in memory, we can transfer the image from the SDRAM to the Zynq SoC's PL for processing using VDMA over a high-performance AXI port.

Should the interface to the camera or the device use a lower-level I/O protocol, we can implement the required interface in the Zynq SoC's PL. These lower-level interfaces typically provide frame and line valid signals along with pixel data. The way these signals are encoded varies, which adds some complexity to the design.

The simplest of these interfaces is a parallel CMOS interface, which provides frame-valid and line-valid signals along with the pixel values in a parallel form, as shown below:

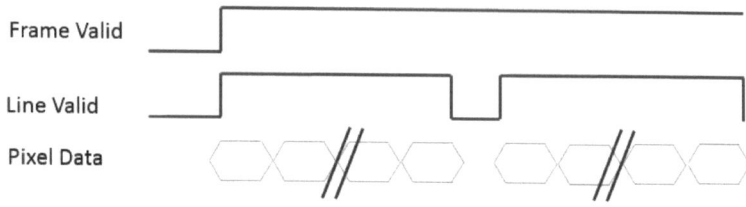

Figure 30 Simple Parallel Video Interface

However, as we increase the frame rate of the image sensor, the use of a parallel CMOS output becomes challenging due to increased signal rates and we usually must use a serialized approach to I/O like Camera Link or LVDS.

Using either Camera Link or serialized LVDS requires that we de-serialize the channels to extract the required information, which involved replicating a parallel structure of pixel value and the frame- and line-valid signals internal to the FPGA as we got directly from the sensor using a parallel interface.

Camera Link comes in three different standards—Base, Medium and Full—providing 2.04, 4.08, and 5.44 Gbps respectively. The base configuration employs four serialized LVDS channels and an LVDS clock running at 85MHz. This interface transfers 24 pixels and 4 framing bits. The Medium and Full versions of the Camera Link interface each introduce another four LVDS Links each so that the Full version has 12 LVDS links.

Image Processing With Xilinx Devices

Camera Link achieves high data rates by serializing data at a rate of 7:1 and transmitting it over 4 LVDS links. The final LVDS link provides the clock, as shown below:

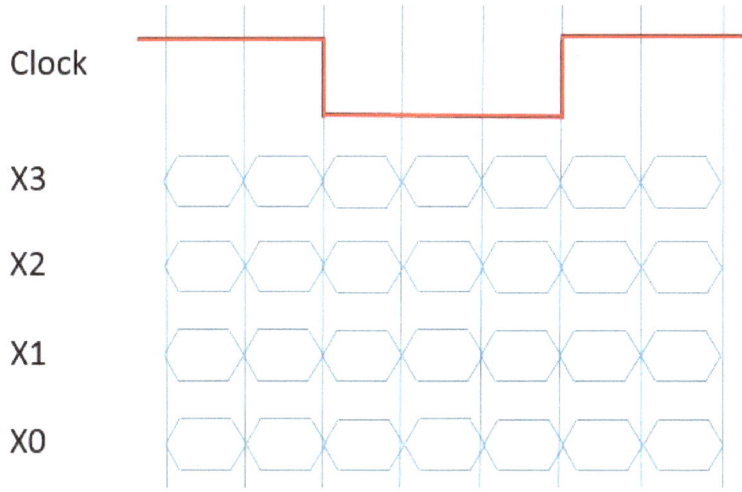

Figure 31 LVDS Camera Link Serialization

When we receive data over a Camera Link interface, we need to de-serialize the five LVDS lines and extract the pixel data in the correct order.

We know the serialisation is 7:1 so we can use one of the MMCMs (mixed-mode clock managers) provided by the Zynq SoC to generate a clock running at 7x the Camera Link clock frequency. However, we still need a framing reference to properly align the received data. Luckily, in the case of Camera Link, we can use the Camera Link clock as the framing reference.

To convert the four LVDS data channels from serial to parallel, we can use the ISERDES2 provided in the Zynq SoC's I/O structure. Using the ISERDES2, we can provide the parallel clock and the higher speed serial clock, and generate a parallel output of as many as 8 bits. (If necessary, we can chain ISERDES2 blocks together for larger parallel outputs.) We need seven outputs for the Camera Link interface, as the serialization is 7:1, so we can de-serialize the interface using only one ISERDES2 for each of the LVDS data channels and the one for the clock.

We use the ISERDES2 from the clock to provide the framing signal. When we have the correct relationship between the received Camera Link clock signal with the generated clock frequency running at 7x the input frequency, the output from this ISERDES2 block will be the pattern "1100011".

We can use a simple state machine to look for this pattern while incrementing or decrementing the phase of the high-speed clock until the correct pattern is detected. Once this pattern is detected, we can than extract the line, frame, and pixel value data from the remaining four ISERDES2 blocks.

Here's a block diagram of the Base Camera Link receiver design based on the above discussion:

Image Processing With Xilinx Devices

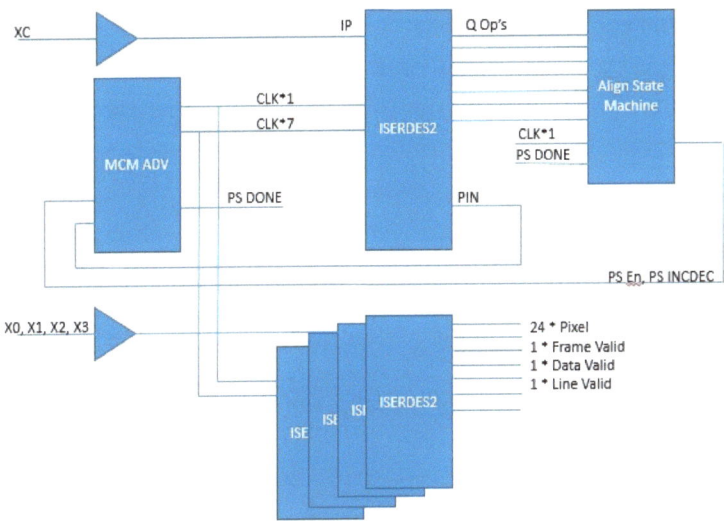

Figure 32 Example Base Camera Link Receiver

We can also take a similar approach to transmitting Camera Link data using an MMCM and OSERDES2 to perform the parallel-to-serial conversion.

While this example uses a Camera Link interface, the same general approach can be used for many serialized I/O applications that provide a signal we can use as a framing reference.

INTERFACING: HDMI

One of the simplest ways to capture or display an image in these applications is using HDMI (High Definition Multimedia Interface). HDMI is a proprietary standard that carries HD digital video and audio data. It is a widely adopted standard supported by many video displays and cameras. Its widespread adoption makes HDMI an ideal interface for our Zynq-based image processing applications.

In this blog, I am going to outline the different options for implementing HDMI in our Zynq design using the different boards we have looked as targets. This exploration will also provide ideas for us when we are designing our own custom hardware.

Figure 33 Arty Z7 HDMI In and Out Example

The several Zynq boards we have used in this series so far support HDMI using one of two methods: an external or internal CODEC.

Image Processing With Xilinx Devices

Board	HDMI IN	HDMI OUT	Comment
Zed Board	FMC - LPC	External CODEC	HDMI IP possible with FMC-HDMI IP
ARTY Z7	Internal CODEC	Internal CODEC	
Pynq	Internal CODEC	Internal CODEC	
Nexys (ARTIX Board, MicroBlaze)	Internal CODEC	Internal CODEC	
MicroZed Embedded Vision Kit	External CODEC	External CODEC	
TySOM	FMC - HPC	External CODEC	Needs HPC FMC IP Card

Table 4 Zynq-based boards with HDMI capabilities

If the board uses an external CODEC, it is fitted with an Analog Devices ADV7511 or ADV7611 for transmission and reception respectively. The external HDMI CODEC interfaces directly with the HDMI connector and generates the TMDS (Transition-Minimized Differential Signalling) signals containing the image and audio data.

The interface between the CODEC and Zynq PL (programmable logic) consists of a I2C bus, pixel-data bus, timing sync signals, and the pixel clock. We route the pixel data, sync signals, and clock directly into the PL. We use the I2C controller in the Zynq PS (processing system) for the I2C interface with the Zynq SoC's I2C IO signals routed via the EMIO to the PL IO.

To ease integration between CODEC and PL, AVNET has developed two IP cores. They are available on the Avnet GitHub. In the image-processing chain, these IP blocks will be located at the very front and end of the chain if you are using them to interface to external CODECs.

The alternate approach is to use an internal CODEC located within the Zynq PL. In this case, the HDMI TMDS signals are routed directly to the PL IO and the CODEC is implemented with programmable logic. To save having to write such complicated CODECs from scratch, Digilent provides two CODEC IP cores. They are available from the Digilent GitHub. Using these cores within the design means the TMDS signals' IO standard within the constraints file is set to TMDS_33 IO.

Note: This IO standard is only available on the High Range (HR) IO banks.

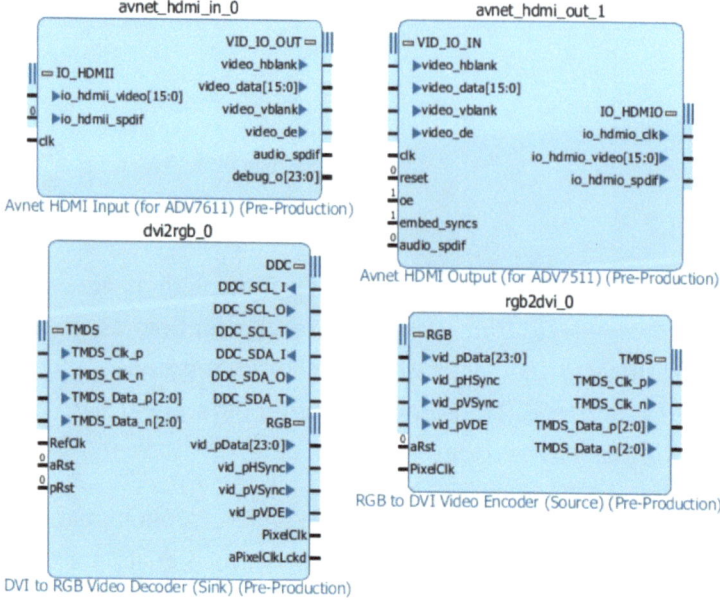

Figure 34 HDMI IP Cores mentioned in the blog

Not every board I have discussed in the MicroZed Chronicles series can both receive and transmit HDMI signals. The ZedBoard and TySOM only provide HDMI output. If we are using one of these boards and the application must receive HDMI signals, we can use the FMC connector with an FMC HDMI input card.

The Digilent FMC-HDMI provides two HDMI inputs with the ability to receive HDMI data using both external and internal CODECs. Of its two inputs, the first uses the ADV7611, while the second equalizes and passes the HDMI Signals through to be decoded directly in the Zynq PL.

Image Processing With Xilinx Devices

Figure 35 Zed board and HDMI FMC

This provides us with the ability to demonstrate how both internal and external CODECs can be implanted on the ZedBoard when using an external CODEC for image transmission.

However first I need to get my soldering iron out to fit a jumper to J18 so that we can set VADJ on the ZedBoard to 3v3 as required for the FMC-HDMI.

We should also remember that while I have predominantly talked about the Zynq SoC here, the same discussion applies to the Zynq UltraScale+ MPSoC, although that device family also incorporates DisplayPort capabilities.

ADV7611 Example

One HDMI receiver discussed is the ADV7611. This device receives three TDMS data streams and converts them into discrete video and audio outputs, which can then be captured and processed. Of course, the ADV7611 is a very capable and somewhat complex device. It requires configuration prior to use. We are going to examine how we can include one within our design.

Figure 36 ZedBoard HDMI Demonstration Configuration

To do this, we need an ADV7611. Helpfully, the FMC HDMI card provides two HDMI inputs, one of which uses an ADV7611. The second equalizes the TMDS data lanes and passes them on directly to the Zynq SoC for decoding.

To demonstrate how we can get this device up and running with our Zynq SoC or Zynq UltraScale+ MPSoC, we will create an example that uses the ZedBoard with the HDMI FMC. For this example, we first need to create a hardware design in Vivado that interfaces with the ADV7611 on the HDMI FMC card. To keep this initial example simple, I will be only receiving the timing signals output by the ADV7611. These signals are:

- Local Locked Clock (LLC) – The pixel clock.
- HSync – Horizontal Sync, indicates the start of a new line.
- VSync – Vertical Sync, indicates the start of a new frame.
- Video Active – indicates that the pixel data is valid (e.g. we are not in a Sync or Blanking period)

Image Processing With Xilinx Devices

This approach uses the VTG's (Video Timing Generator's) detector to receive the sync signals and identify the received video's timing parameters and video mode. Once the ADV7611 correctly identifies the video mode, we have configured correctly. It is then a simple step to connect the received pixel data to a Video-to-AXIS IP block and use VDMA to write the received video frames into DDR memory for further processing.

For this example, we need the following IP blocks:

- VTC (Video Timing Controller) – Configured for detection and to receive sync signals only.
- ILA – Connected to the sync signals so that we can see that they are toggling correctly—to aid debugging and commissioning.
- Constant – Set to a constant 1 to enable the clock and detector enables.

The resulting block diagram appears below. The eagle-eyed will also notice the addition both a GPIO output and I2C bus from the processor system. We need these to control and configure the ADV7611.

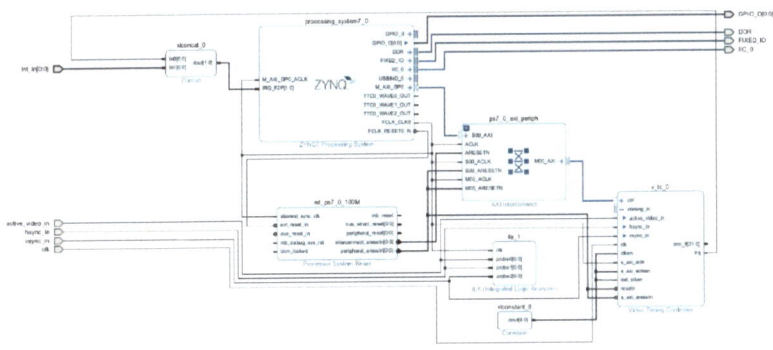

Figure 37 Simple Architecture to detect the video type

Following power up, the ADV7611 generates no sync signals or video. We must first configure the device, which requires the use of an I2C bus. We therefore need to enable one of the two I2C controllers within the Zynq PS and route the IO to the EMIO so that we can then route the I2C signals (SDA and SCL) to the correct pins on the FMC connector. The ADV7611 is a complex device to configure with multiple I2C addresses that address different internal functions within the device. EDID and High-bandwidth Digital Content Protection (HDCP), for example.

We also need to be able to reset the ADV7611 following the application of power to the ZedBoard and FMC HDMI. We use a PS GPIO pin, output via the EMIO, to do this. Using a controllable I/O pin for this function allows the application software to reset of the device each time we run the program. This capability is also helpful when debugging the software application to ensure that we start from a fresh reset each time the program runs—a procedure that prevents previous configurations form affecting the next.

With the block diagram completed, all that remains is to build the design with the location constraints (identified below) to connect to the correct pins on the FMC connector for the ADV7611.

```
set_property PACKAGE_PIN M19 [get_ports {clk}];           # "FMC-LA00_CC_P"
set_property PACKAGE_PIN G16 [get_ports {hsync_in}];      # "FMC-LA19_N"
set_property PACKAGE_PIN E20 [get_ports {iic_0_scl_io}];  # "FMC-LA21_N"
set_property PACKAGE_PIN F19 [get_ports {active_video_in}]; # "FMC-LA22_N"
set_property PACKAGE_PIN G19 [get_ports {vsync_in}];      # "FMC-LA22_P"
set_property PACKAGE_PIN E15 [get_ports {GPIO_O}];        # "FMC-LA23_P"
set_property PACKAGE_PIN C22 [get_ports {Int_in}];        # "FMC-LA25_N"
set_property PACKAGE_PIN D22 [get_ports {iic_0_sda_io}];  # "FMC-LA25_P"

set_property IOSTANDARD LVCMOS18 [get_ports -of_objects [get_iobanks 34]];
set_property IOSTANDARD LVCMOS18 [get_ports -of_objects [get_iobanks 35]];
```

Figure 38 Vivado Constraints for the ADV7611 Design

With the bit file generated, we are now able to create software that configures the ADV7611 Low-Power HDMI Receiver chip and the Zynq SoC's VTC (Video Timing Controller). If we do this correctly, the VTC will then be able to report the input video mode.

Image Processing With Xilinx Devices

To be able to receive and detect the video mode, the software must perform the following steps:

1) Initialize and configure the Zynq SoC's I2C controller for master operation at 100KHz
2) Initialize and configure the VTC
3) Configure the ADV7611
4) Sample the VTC once a second, reporting the detected video mode

Figure 39 ZedBoard, FMC HDMI, and the PYNQ dev board connected for testing

Configuring the I2C and VTC is very simple. We have done both several times throughout this series (See these links: I2C, VTC.) Configuring the ADV7611 is more complicated and is performed using I2C. This is where this example gets a little complicated as the ADV7611 uses eight internal I2C slave addresses to configure different sub functions.

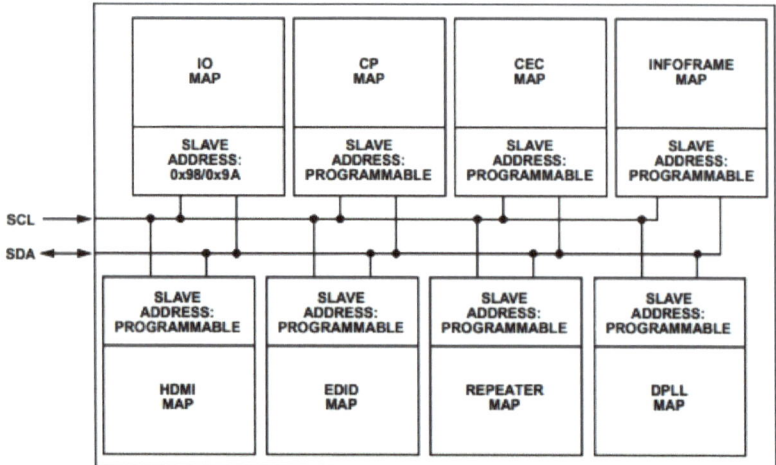

Figure 65. ADV7611 Register Map Access through Main I²C Port

Figure 40 ADV7611 I2C Architecture

To reduce address-contention issues, seven of these addresses are user configurable. Only the IO Map has a fixed default address.

I2C addressing uses 7 bits. However, the ADV7611 documentation specifies 8-bit addresses, which includes a Read/Write bit. If we do not understand the translation between these 7- and 8-bit addresses, we will experience addressing issues because the Read/Write bit is set or cleared depending on the function we call from XIICPS.h.

The picture below shows the conversion from 8-bit to 7-bit format. The simplest method is to shift the 8-bit address one place to the right.

Bit 7	Bit 6	Bit 5	Bit 4	Bit 3	Bit 2	Bit 1	Bit 0	ADDR	Comment
Address Bit MSB						Address Bit LSB	R/W		
1	0	0	1	1	0	0	0	0x98	8 bit format
1	0	0	1	1	0	0	NA	0x4C	7 Bit format as used

Table 5 I2C addressing explained

We need to create a header file containing the commands to configure each of the eight ADV7611's sub functions.

Image Processing With Xilinx Devices

This raises the question of where to obtain the information to configure the ADV7611 device. Rather helpfully, the Analog Devices engineer zone, provides several resources including a recommended registers settings guide and several pre-tested scripts that you can download and use to configure the device for most use cases. All we need to do is select the desired use case and incorporate the commands into our header file.

One thing we must be very careful with is that the first command issued to the AD7611 must be an I2C reset command. You may see a NACK on the I2C bus in response to this command as the reset asserts very quickly. We also need to wait an appropriate period after issuing the reset command before continuing to load commands. In this example, I decided to wait the same time as following a hard reset, which the data sheet specifies as 5msec.

Once 5msec has elapsed following the reset, we can continue loading configuration data, which includes the Extended Display Identification Data (EDID) table. The EDID identifies to the source the capabilities of the display. Without a valid EDID table, the HDMI source will not start transmitting data.

Having properly configured the ADV7611, we may want to read back registers to ensure that it is properly configured or to access the device's status. To do this successfully, we need to perform what is known as a I2C repeat start in the transaction following the initial I2C write. A repeat start is used when a master issues a write command and then wants to read back the result immediately. Issuing the repeat start prevents another device from interrupting the sequence.

We can configure the I2C controller to issue repeat starts between write and read operations within our software application by using the function call XIicPs_SetOptions(&Iic,XIICPS_REP_START_OPTION). Once we have completed the transaction we need to clear the repeat start option using the XIicPs_ClearOptions(&Iic,XIICPS_REP_START_OPTION) function call. Otherwise we may have issues with communication.

Once configured, the ADV7611 starts free running. It will generate HDMI Frames even with no source connected. The VTC will receive these input frames, lock to them and determine the video mode. We can obtain both the timing parameters and video mode by using the VTC API. The video modes that can be detected are:

```
#define XVTC_VMODE_720P        1   /**< Video mode 720P */
#define XVTC_VMODE_1080P       2   /**< Video mode 1080P */
#define XVTC_VMODE_480P        3   /**< Video mode 480P */
#define XVTC_VMODE_576P        4   /**< Video mode 576P */
#define XVTC_VMODE_VGA         5   /**< Video mode VGA */
#define XVTC_VMODE_SVGA        6   /**< Video mode SVGA */
#define XVTC_VMODE_XGA         7   /**< Video mode XGA */
#define XVTC_VMODE_SXGA        8   /**< Video mode SXGA */
#define XVTC_VMODE_WXGAPLUS    9   /**< Video mode WXGAPlus */
#define XVTC_VMODE_WSXGAPLUS   10  /**< Video mode WSXGAPlus */
#define XVTC_VMODE_1080I       100 /**< Video mode 1080I */
#define XVTC_VMODE_NTSC        101 /**< Video mode NTSC */
#define XVTC_VMODE_PAL         102 /**< Video mode PAL */
```

Figure 41 Inbuilt Video detection modes

Initially in its free-running mode, the ADV7611 outputs video in 480x640 pixel format. Checking the VTC registers, it is also possible to observe that the detector has locked with the incoming sync signals and has detected the mode correctly, as shown in the image below:

Image Processing With Xilinx Devices

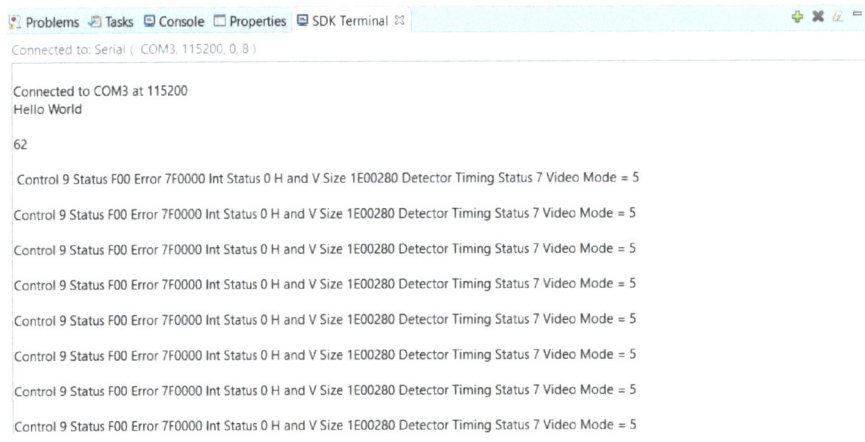

Figure 42 ADV7611 free running VGA mode

With the free-running mode functioning properly, the next step is to stimulate the FMC HDMI with different resolutions to ensure that they are correctly detected.

To test the application, we will use a PYNQ Dev Board. The PYNQ is ideal for this application because it is easily configured for different HDMI video standards using just a few lines of Python, as shown below. The only downside is the PYNQ board does not generate fully compliant 1080P video timing.

```
Control 9 Status F00 Error 7F0000 Int Status 0 H and V Size 2580320 Detector Timing Status 7 Video Mode = 6
Control 9 Status F00 Error 7F0000 Int Status 0 H and V Size 2580320 Detector Timing Status 7 Video Mode = 6
Control 9 Status F00 Error 7F0000 Int Status 0 H and V Size 2580320 Detector Timing Status 7 Video Mode = 6
```

Figure 43 SVGA video outputting 800 pixels by 600 lines @ 60Hz

```
Control 9 Status F00 Error 7F0000 Int Status 0 H and V Size 2D00500 Detector Timing Status 7 Video Mode = 1
Control 9 Status F00 Error 7F0000 Int Status 0 H and V Size 2D00500 Detector Timing Status 7 Video Mode = 1
Control 9 Status F00 Error 7F0000 Int Status 0 H and V Size 2D00500 Detector Timing Status 7 Video Mode = 1
```

Figure 44 720P video outputting 1280 pixels by 720 Lines @ 60 Hz

Control 9 Status F00 Error 7F0000 Int Status 0 H and V Size 40004FF Detector Timing Status 7 Video Mode = 8

Control 9 Status F00 Error 7F0000 Int Status 0 H and V Size 40004FF Detector Timing Status 7 Video Mode = 8

Control 9 Status F00 Error 7F0000 Int Status 0 H and V Size 40004FF Detector Timing Status 7 Video Mode = 8

Control 9 Status F00 Error 7F0000 Int Status 0 H and V Size 40004FF Detector Timing Status 7 Video Mode = 8

Figure 45 SXGA video outputting 1280 pixels by 1024 lines @ 60Hz

Having performed these tests, it is clear the ADV7611 on the FMC HDMI is working as required and is receiving and decoding different HDMI resolutions correctly. At the same time, the VTC is correctly detecting the video mode, enabling us to capture video data on our Zynq SoC or Zynq UltraScale+ MPSoC systems for further processing.

The FMC HDMI has another method of receiving HDMI that equalizes the channel and passes it through to the Zynq SoC's or Zynq UltraScale+ MPSoC's PL for decoding. I will create an example design based upon that input over the next few weeks.

Note that we can also use this same approach with a MicroBlaze soft processor core instantiated in a Xilinx FPGA.

Direct Decoding Example

In this example we are going to look at using internal IP cores to receive HDMI images in conjunction with the Analog Devices [AD8195](AD8195) HDMI buffer, which equalizes the line. Equalization is critical when using long HDMI cables.

Image Processing With Xilinx Devices

Figure 46 Nexys board, FMC HDMI and the Digilent PYNQ-Z1

To do this I will be using the <u>Digilent FMC HDMI</u> card, which provisions one of its channels with an AD8195. The AD8195I on the FMC HDMI card needs a 3v3 supply, which is not available on the ZedBoard unless I break out my soldering iron. Instead, I broke out my <u>Digilent Nexys Video trainer board</u>, which comes fitted with an Artix-7 FPGA and an FMC connector. This board has built-in support for HDMI RX and TX but the HDMI RX path on this board supports only 1m of HDMI cable while the AD8195 on the FMC HDMI card supports cable runs of up to 20m—far more useful in many distributed applications. So we'll add the FMC HDMI card.

First, I instantiated a MicroBlaze soft microprocessor system in the Nexys Video card's Artix-7 FPGA to control the simple image-processing chain needed for this example. Of course, you can implement the same approach to the logic design that I outline here using a Xilinx Zynq SoC or Zynq UltraScale+ MPSoC. The Zynq PS simply replaces the MicroBlaze.

The hardware design we need to build this system is:

- MicroBlaze controller with local memory, AXI UART, MicroBlaze Interrupt controller, and DDR Memory Interface Generator.
- [DVI2RGB IP](#) core to receive the HDMI signals and convert them to a parallel video format.
- Video Timing Controller, configured for detection.
- ILA connected between the VTC and the DVI2RBG cores, used for verification.
- Clock Wizard used to generate a 200MHz clock, which supplies the DDR MIG and DVI2RGB cores. All other cores are clocked by the MIG UI clock output.
- Two 3-bit GPIO modules. The first module will set the VADJ to 3v3 on the HDMI FMC. The second module enables the ADV8195 and provides the hot-plug detection.

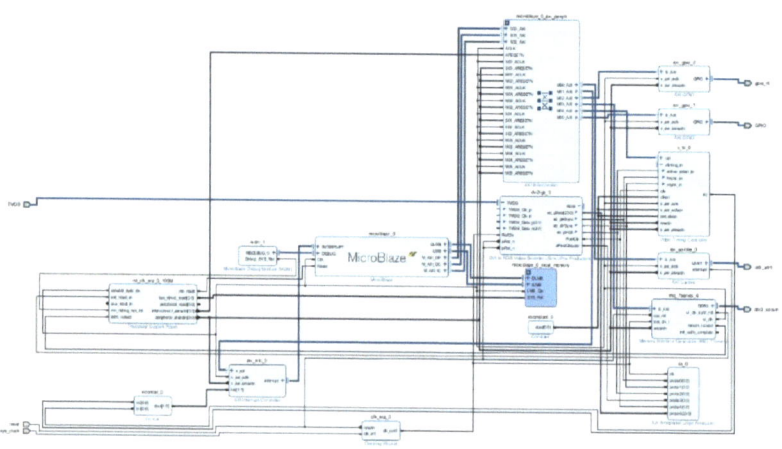

Figure 47 Architecture of the design

The final step in this hardware build is to map the interface pins from the AD8195 to the FPGA's I/O pins through the FMC connector. We'll use the TMDS_33 SelectIO standard for the HDMI clock and data lanes.

Once the hardware is built, we need to write some simple software to perform the following:

Image Processing With Xilinx Devices

- Disable the VADJ regulator using pin 2 on the first GPIO port.
- Set the desired output voltage on VADJ using pins 0 & 1 on the first GPIO port.
- Enable the VADJ regulator using pin 2 on the first GPIO port.
- Enable the AD8195 using pin 0 on the second GPIO port.
- Enable pre- equalization using pin 1 on the second GPIO port.
- Assert the Hot-Plug Detection signal using pin 2 on the second GPIO port.
- Read the registers within the VTC to report the modes and status of the video received.

To test this system, I used a Digilent PYNQ-Z1 board to generate different video modes. The first step in verifying that this interface is working is to use the ILA to check that the pixel clock is received and that its DLL is locked, along with generating horizontal and vertical sync signals and the correct pixel values.

Provided the sync signals and pixel clock are present, the VTC will be able to detect and classify the video mode. The application software will then report the detected mode via the terminal window.

MicroZed Chronicles Part 220

www.adiuvoengineering.com

H and V Size 2580320 Detector Timing Status 7 Video Mode = 6

H and V Size 2580320 Detector Timing Status 7 Video Mode = 6

H and V Size 2580320 Detector Timing Status 7 Video Mode = 6

Figure 48 Software running on the Nexys Video detecting SVGA mode (600 pixels by 800 lines)

With the correct video mode being detected by the VTC, we can now configure a VDMA write channel to move the image from the logic into a DDR frame buffer.

USING HIGH LEVEL SYNTHESIS

So what is wrong with plain, old HDL design? Why do we need a whole new design-capture paradigm? As devices get larger and as designs get more complicated, we need new to use new development methodologies to achieve the required results while keeping development time reasonable. This has always been the case for Programmable Logic.

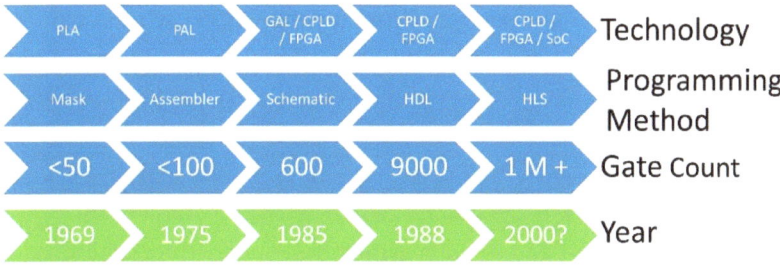

Figure 49 Overview of Programmable Logic Development over the last 40 plus years.

As the marketers like to say, HLS raises the level of design abstraction and that higher level allows us to create complex designs more quickly. The marketers like to say this, and in this instance they happen to be correct. HLS has been on the verge of breaking into mainstream development for a number of years. (Certainly I remember a demo when I was a graduate in 2000 of HandelC.) However, recent development-tool introductions including the Vivado HLx releases have pushed HLS firmly into mainstream design. I now use it a lot and like it.

That said there is still a place for HDL modules where handcrafted hardware design is needed to achieve maximum performance. The key is knowing what tool we as engineers should use for the task at hand so that we can deliver designs while meeting quality, time and budget goals.

The HLS environment within Vivado HLx should be very familiar to you if you are developing with the Xilinx Zynq SoC or Zynq UltraScale+ MPSoC using SDK or SDSoC. Vivado HLx and SDSoC are both based on the Eclipse platform.

We can use Vivado HLS to directly generate HDL from C, C++, or SystemC descriptions. The code for the function we want to compile with HLS must stand alone and must be contained within its own source file. (We will examine more HLS constraints in future blogs). Another big advantage of using HLS is that you also define the test bench in C, which allows for much quicker verification. What is really cool is we can use this high-level test bench to verify the HDL module compiled from the C/C++/SystemC code.

The Vivado HLS menu bar allows us to implement all of the steps we need. The menu bar shows C Simulation, Synthesis, Co-Simulation, Export RTL, Open Report and View Waveform—roughly the order in which they are used.

Figure 50 Vivado HLS menu bar

The Vivado HLS design flow is very simple, as we will see with a simple example (which adds two integers together). This flow is outlined below:

1. Write the C function we wish to compile with HLS. Add this source file to the source directory within Vivado HLS. (See add_ex.c and add_ex.h in the GitHub repository.)

2. Write a C test bench that tests the C function we are going to compile with Vivado HLS. Put this test-bench file in the test bench directory within Vivado HLS. (See add_tb.c on the GitHub repository.)

Image Processing With Xilinx Devices

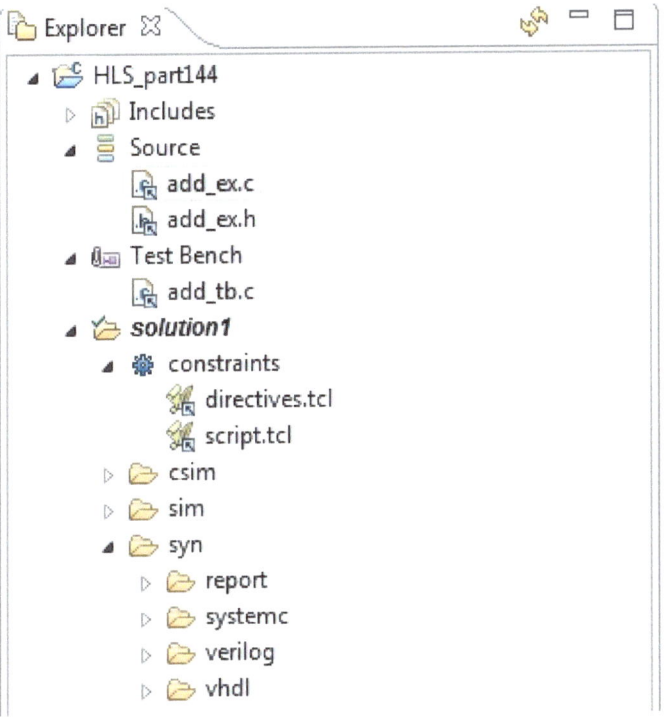

Figure 51 Source files and test bench files within Vivado HLS

3. Run a C simulation within Vivado HLS. This simulation allows us to verify that our C function works correctly. We want to know that the function works in C before we move on to synthesis and Co-Simulation. Within this test bench, we can use the standard techniques that we would in most C test harnesses. As this test bench runs within the HLS debugging environment, we can set breakpoints, examine memory, and step through the program should there be issues with either source or test bench code.

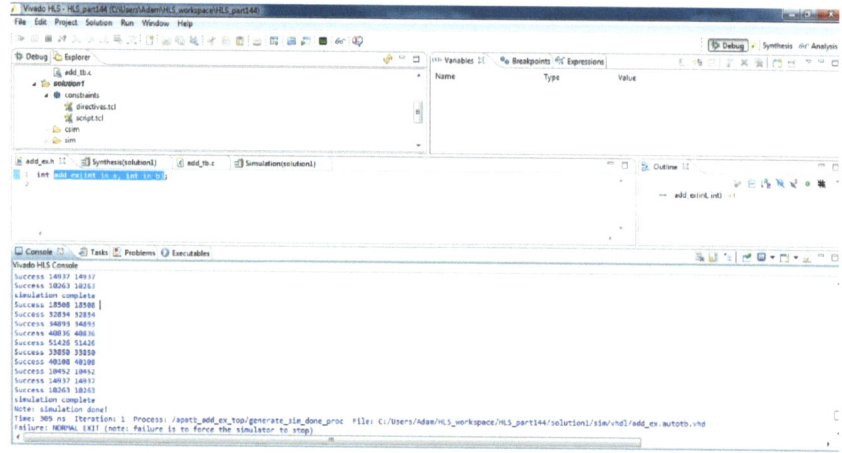

Figure 52 C Simulation and Reported Results in the Console

4. If we are happy with the results of the C simulation, we can progress to synthesizing the module to generate HDL code for our design. This step will also produce a synthesis report that includes resource estimates and interface parameters.

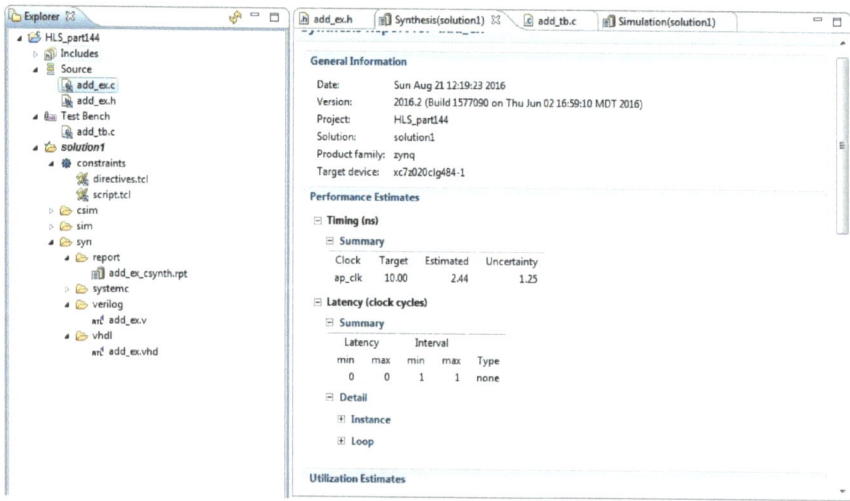

Figure 53 HLS Synthesis Report & Output files

Image Processing With Xilinx Devices

Utilization Estimates

Summary

Name	BRAM_18K	DSP48E	FF	LUT
DSP	-	-	-	-
Expression	-	-	0	32
FIFO	-	-	-	-
Instance	-	-	-	-
Memory	-	-	-	-
Multiplexer	-	-	-	-
Register	-	-	-	-
Total	0	0	0	32
Available	280	220	106400	53200
Utilization (%)	0	0	0	~0

Figure 54 Report Utilization Estimate

5. Once we have the synthesised output, we can run co-simulation, which will apply the C test bench to the generated HDL using the selected simulator (Verilog or VHDL) and will report on the pass/fail criteria we set. This will also produce a co-simulation report and, even better, we can capture waveforms from the simulation and open them in the simulation tool for later inspection.

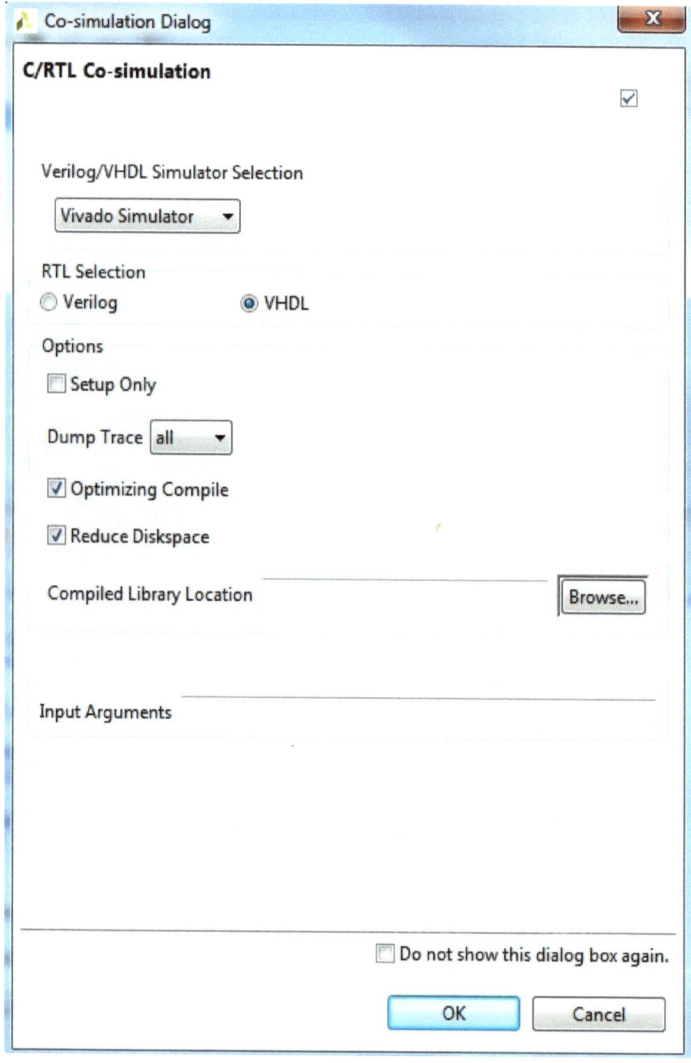

Figure 55 Co-simulation dialog box allowing us to select Verilog or VHDL HDL

Cosimulation Report for 'add_ex'

Result

RTL	Status	Latency			Interval		
		min	avg	max	min	avg	max
VHDL	Pass	0	0	0	0	0	0
Verilog	NA	NA	NA	NA	NA	NA	NA

Export the report(.html) using the Export Wizard

Figure 56 Co-simulation Report

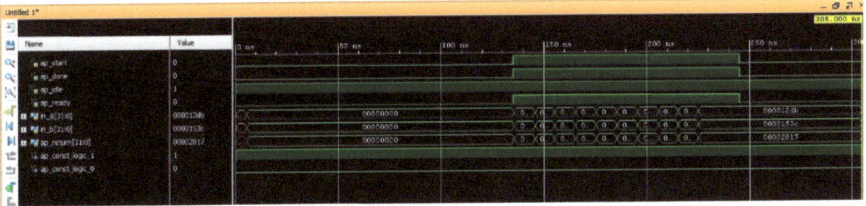

Figure 57 Screen Shot of Co-simulation waveform results

6. The final development step is to export the resulting HDL to the Vivado IP library so that we can use it in our designs.

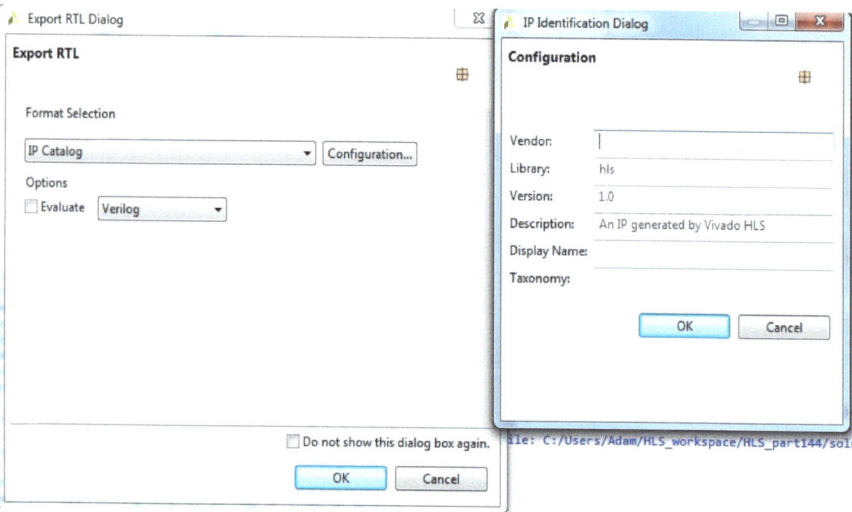

Figure 58 RTL Export Dialog

Overall the Vivado HLS flow is very simple and tightly integrated.

To be able to synthesise C code using Vivado HLS we must follow some simple rules. The rules are:

- The function must be self-contained; that is it must contain the entire design to be implemented.
- You cannot use system calls or dynamic memory allocations.
- You must use fixed or bounded constructs
- The function must be unambiguous in its use of the bounds.

With the ground rules set for creating the function, I now want to look at how we define the interface to our function in the final RTL module. Generally speaking, there are two interface types to consider:

- Block-Level Interface – This is the default for a generic interface that provides clock and reset inputs along with start, ready, idle, and done handshaking signals.

Image Processing With Xilinx Devices

- Port-Level Interface – This interface implements a more complete I/O protocol on the data ports such as AXI, AXI streaming, or AXI Lite.

We can use the simple Block-Level interface if we want, without the need for a Port-Level interface. If this is the case, we will need to ensure that the data on the block's inputs is stable for the I/O operation.

Looking at the example from last week, the synthesised results produced a very simple entity that implemented a Block-Level Interface.

```
 8 library IEEE;
 9 use IEEE.std_logic_1164.all;
10 use IEEE.numeric_std.all;
11
12 entity add_ex is
13 port (
14     ap_start : IN STD_LOGIC;
15     ap_done : OUT STD_LOGIC;
16     ap_idle : OUT STD_LOGIC;
17     ap_ready : OUT STD_LOGIC;
18     in_a : IN STD_LOGIC_VECTOR (31 downto 0);
19     in_b : IN STD_LOGIC_VECTOR (31 downto 0);
20     ap_return : OUT STD_LOGIC_VECTOR (31 downto 0) );
21 end;
22
```

Figure 59 Block-Level Interface from the example above

There are three options we can apply when we use the Block-Level Interface. The first, AP_NONE, will generate no interfaces beyond the ones defined within the module. You can use this option for very simple modules. However, should you choose this option, you will not be able to use Co-Simulation to verify your design.

```
 8 library IEEE;
 9 use IEEE.std_logic_1164.all;
10 use IEEE.numeric_std.all;
11
12 entity add_ex is
13 port (
14     in_a : IN STD_LOGIC_VECTOR (31 downto 0);
15     in_b : IN STD_LOGIC_VECTOR (31 downto 0);
16     ap_return : OUT STD_LOGIC_VECTOR (31 downto 0) );
17 end;
18
```

Figure 60 HLS results with AP_NONE set

The default option for the Block-Level interface is AP_CNTRL_HS, which provides the required clock, reset, and handshaking ports needed to create a handshake for the block, as discussed above. The final option, AP_CNTRL_CHAIN, is very similar in that it provides the same handshaking signals as AP_CNTRL_HS but adds the ability to chain HLS blocks together by adding a continue signal, which can be used to stall an upstream block if necessary.

Depending on the structure we create with our HLS module, there are a number of other options we can apply to the Block-Level interface such as "MEMORY" or "FIFO," for example.

We can control these options using pragmas to define the block's I/O architecture or we can use the directives window, which opens a dialog allowing us to select the I/O architecture that we want to use.

By highlighting the top-level function in the directives window, we can select the Block-Level interface or the AXI interface. We can also do this for the inputs by selecting the required standards.

Image Processing With Xilinx Devices

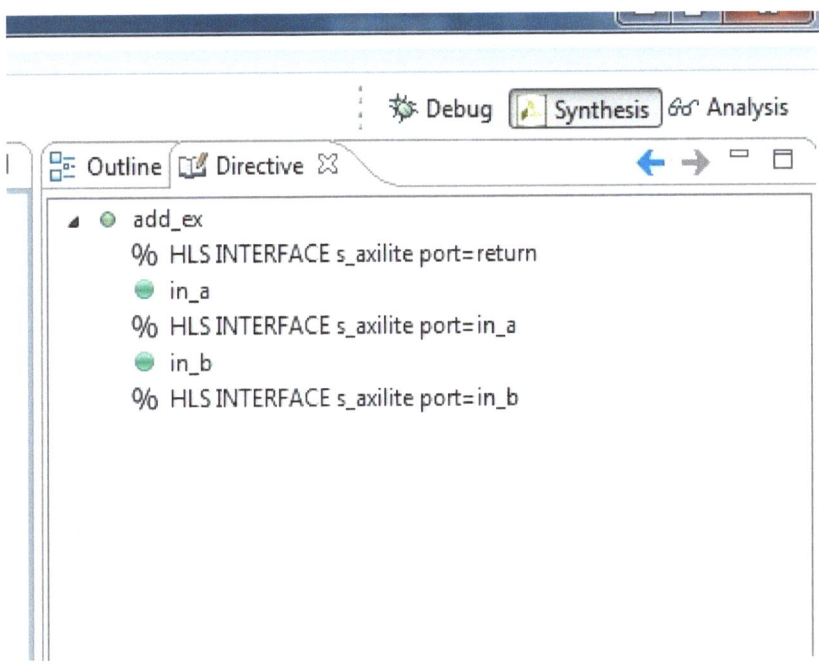

Figure 61 Directives Window – Set to use AXI Lite

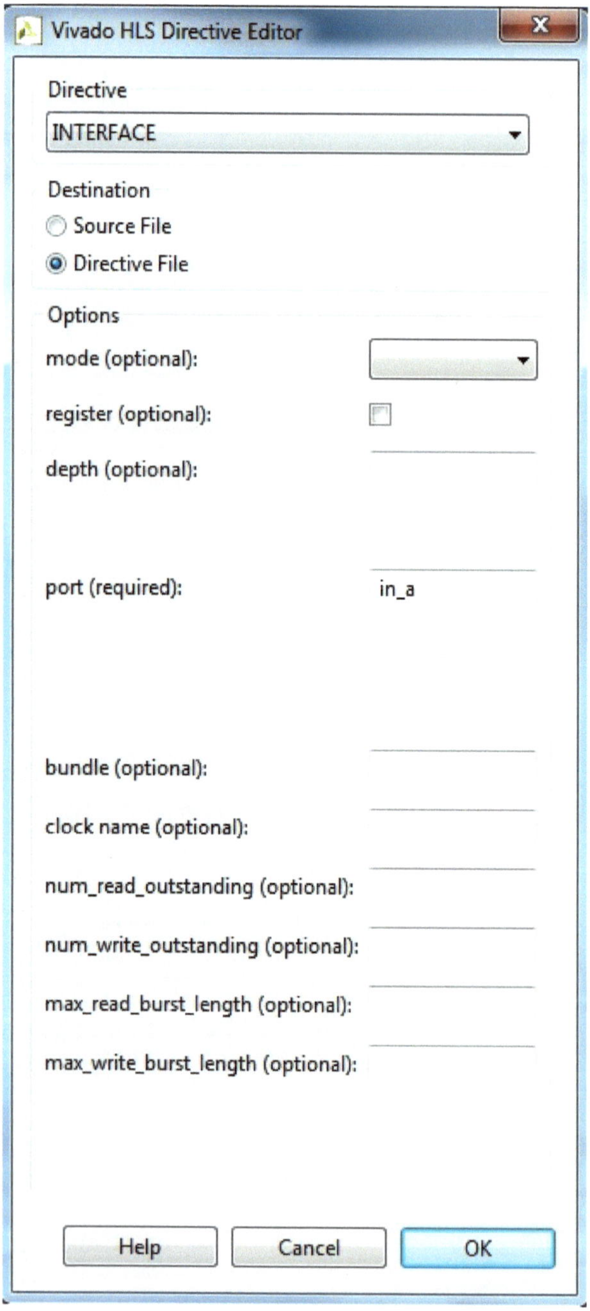

Figure 62 Directives Insertion Window for Input A

Image Processing With Xilinx Devices

If we decide to use a pragma-based approach to define out interfaces, we can do so as shown below which creates AXI Streaming and AXI Lite interfaces.

```
void image_filter(AXI_STREAM& video_in, AXI_STREAM& video_out, int rows, int cols)
{
#pragma HLS INTERFACE axis port=video_in bundle=INPUT_STREAM
#pragma HLS INTERFACE axis port=video_out bundle=OUTPUT_STREAM
#pragma HLS INTERFACE s_axilite port=rows bundle=CONTROL_BUS offset=0x14
#pragma HLS INTERFACE s_axilite port=cols bundle=CONTROL_BUS offset=0x1C
#pragma HLS INTERFACE s_axilite port=return bundle=CONTROL_BUS
#pragma HLS INTERFACE ap_stable port=rows
#pragma HLS INTERFACE ap_stable port=cols
```

Figure 63 Defining an Interface using pragmas

Setting our simple example from last week to provide an AXI Lite interface using the directives window, you can see the resultant interface below:

```vhdl
 8 library IEEE;
 9 use IEEE.std_logic_1164.all;
10 use IEEE.numeric_std.all;
11
12 entity add_ex is
13 generic (
14     C_S_AXI_AXILITES_ADDR_WIDTH : INTEGER := 6;
15     C_S_AXI_AXILITES_DATA_WIDTH : INTEGER := 32 );
16 port (
17     s_axi_AXILiteS_AWVALID : IN STD_LOGIC;
18     s_axi_AXILiteS_AWREADY : OUT STD_LOGIC;
19     s_axi_AXILiteS_AWADDR : IN STD_LOGIC_VECTOR (C_S_AXI_AXILITES_ADDR_WIDTH-1 downto 0);
20     s_axi_AXILiteS_WVALID : IN STD_LOGIC;
21     s_axi_AXILiteS_WREADY : OUT STD_LOGIC;
22     s_axi_AXILiteS_WDATA : IN STD_LOGIC_VECTOR (C_S_AXI_AXILITES_DATA_WIDTH-1 downto 0);
23     s_axi_AXILiteS_WSTRB : IN STD_LOGIC_VECTOR (C_S_AXI_AXILITES_DATA_WIDTH/8-1 downto 0);
24     s_axi_AXILiteS_ARVALID : IN STD_LOGIC;
25     s_axi_AXILiteS_ARREADY : OUT STD_LOGIC;
26     s_axi_AXILiteS_ARADDR : IN STD_LOGIC_VECTOR (C_S_AXI_AXILITES_ADDR_WIDTH-1 downto 0);
27     s_axi_AXILiteS_RVALID : OUT STD_LOGIC;
28     s_axi_AXILiteS_RREADY : IN STD_LOGIC;
29     s_axi_AXILiteS_RDATA : OUT STD_LOGIC_VECTOR (C_S_AXI_AXILITES_DATA_WIDTH-1 downto 0);
30     s_axi_AXILiteS_RRESP : OUT STD_LOGIC_VECTOR (1 downto 0);
31     s_axi_AXILiteS_BVALID : OUT STD_LOGIC;
32     s_axi_AXILiteS_BREADY : IN STD_LOGIC;
33     s_axi_AXILiteS_BRESP : OUT STD_LOGIC_VECTOR (1 downto 0);
34     ap_clk : IN STD_LOGIC;
35     ap_rst_n : IN STD_LOGIC;
36     interrupt : OUT STD_LOGIC );
37 end;
```

Figure 64 AXI Slave Interface for example above

Depending upon the type of AXI interface (indeed any interface) we wish to implement, we need to consider if we are passing the arguments to the function by either value or reference. Here's a figure that summarizes the argument-passing options for different argument types:

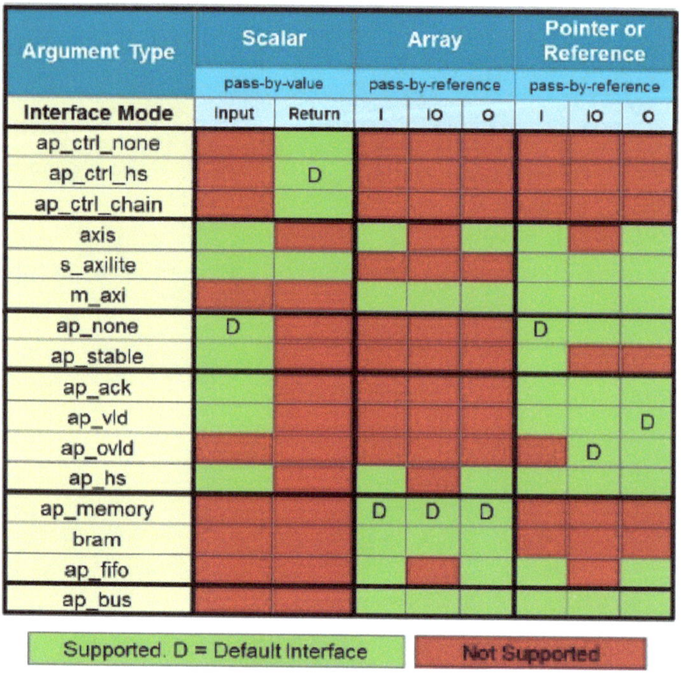

Argument Type	Scalar		Array			Pointer or Reference		
	pass-by-value		pass-by-reference			pass-by-reference		
Interface Mode	Input	Return	I	IO	O	I	IO	O
ap_ctrl_none								
ap_ctrl_hs		D						
ap_ctrl_chain								
axis								
s_axilite								
m_axi								
ap_none	D					D		
ap_stable								
ap_ack								
ap_vld								D
ap_ovld							D	
ap_hs								
ap_memory			D	D	D			
bram								
ap_fifo								
ap_bus								

Supported. D = Default Interface   Not Supported

Figure 65 Interface type and Function passing style for different interfaces

Having covered how we can control our resultant module interface.

But what about the functionality we desire within the module itself, like with a traditional HDL design we could design it all ourselves from scratch. However, as we want to deliver on time quality and cost to our customers, when we develop traditional HDL designs we normally use IP blocks where possible and create new modules only which are of added value.

Image Processing With Xilinx Devices

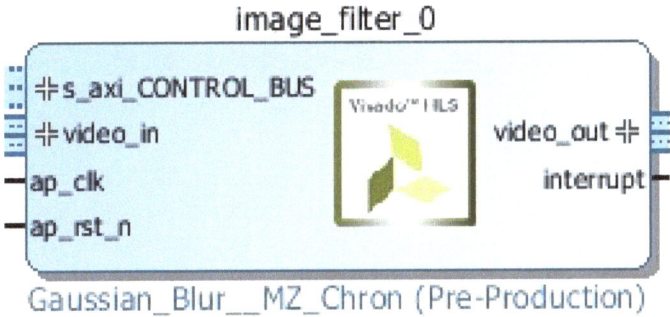

Figure 66 The IP Module we will create

When it comes to HLS we need to follow a similar approach, opposed to writing everything from scratch. Thankfully Vivado HLS provides a number of libraries allowing us to develop our values added areas as before and not worry about the more commonplace aspects. Only this time we get the benefits of working with a higher level of abstraction.

With our HLS installation we get six libraries as standard that provide a range of functions for us

- Math Library – Provides synthesisable implementations of the standard maths libraries.
- Video Library – Provides Video processing libraries similar to OpenCV
- IP Library – Provides IP libraries for implementing FFT, FIR and Shift Register Look Up Table functions.
- Linear Algebra Library – Provides a library of commonly used Linear Algebra functions
- Stream Library – Provides structures for modelling Streaming data interfaces
- Arbitrary Precision Data Types Library – Provides support for none power of 2 arbitrary lengths for signed and unsigned integers. This allows for more efficient use of the resources of the FPGA.

It is one of these libraries that we are going to explore in depth next and that is the Video Library, this library allows us to accelerate the development of an image processing pipeline. To demonstrate the functionality of the IP blocks in the programmable logic, we will be using the Avnet Embedded Vision Kit.

The Video HLS libraries include a range of functions to help us test and implement the embedded vision systems this includes OpenCV libraries.

When it comes to implementation in a FPGA using HLS these libraries provide similar functions to those provided by OpenCV. However, it is not possible to perform HLS synthesis directly on OpenCV libraries as they are not optimised for implementation in FPGA fabric due to the use of dynamic memory allocation.

The HLS environment provides two libraries we can use to develop our EV application

- hls_video – This library provides EV functions and data structures, these elements can be synthesised.
- hls_opencv – This library includes the pre compiled OpenCV functions along with special support functions required to interface with the IP module. This library is intended for use by the test bench and as such in not synthesisable.

Using our HLS environment, we can if we wish still develop at a high level initially using the hls_opencv library to prototype the algorithm. This allows us to quickly and easily create the system model of the algorithm within our HLS environment and verify the results meet the requirements.

Our next step is to understand how we store an image and the subtle difference between OpenCV and the HLS Video Libraries.

Image Processing With Xilinx Devices

Figure 67 Different types of edge detection (Original, Laplacian of Gaussian, Canny and Sobel)

The most basic of OpenCV elements is the cv::mat class, which defines the image size in X and Y and pixel information (e.g. the number of bits within pixel), if the pixel data is signed or unsigned, and how many channels make up a pixel. This class creates the basis for how we store and manipulate images when we use OpenCV.

Within the HLS library there is a similar construct: the hls::mat. The library also providesa number of functions that enable conversion of the hls::mat class to and from HLS streaming. This is the standard interface we use when creating image-processing pipelines. One major difference between the cv::mat and the hls::mat classes is that the hls::mat class is defined as a stream of pixels as opposed to the cv::mat definition, which is a block of memory. This difference means that we do not have random access to pixels using hls::mat.

A simple example that demonstrates how we can use these libraries is to perform a simple Gaussian Blur of an image. The filter will use AXI Streaming interfaces to input and output the image data stream.

Gaussian blurring is typically applied to an image prior to many edge-detection embedded-vision algorithms that reduce noise within the image like Sobel or Canny.

The first step is to create the HLS structures we need within a header file so that both the module to be synthesised and the test bench can use them. These type definitions are:

1. HLS streaming interfaces: this makes using the conversion to and from AXI streams within the test bench easier.
   typedef hls::stream<ap_axiu<16,1,1,1> >   AXI_STREAM;

2. HLS mat type: if we are using RGB and YUV we will need to define different types
   typedef hls::Mat<MAX_HEIGHT, MAX_WIDTH, HLS_8UC2> YUV_IMAGE;
   typedef hls::Mat<MAX_HEIGHT, MAX_WIDTH, HLS_8UC3> RGB_IMAGE;

With the basics defined we are then in a position to generate the module we wish to synthesise and the test bench to check it is functioning.

Starting with the module we wish to synthesise, the video input and output from the module will use the previously defined AXI_STREAM type definition. While the size of the image in rows and columns will be supplied over an AXI-Lite interface, we can also use this interface if we want to provide the ability to enable or disable the filter.

Implementing the function we want is very simple. We need to convert the input video from an AXI Stream into an hls::mat, apply our filter, and then convert the output hls::mat back to an AXI Stream.

Image Processing With Xilinx Devices

```
void image_filter(AXI_STREAM& video_in, AXI_STREAM& video_out, int rows, int cols) {
    //Create AXI streaming interfaces for the core
#pragma HLS INTERFACE axis port=video_in bundle=INPUT_STREAM
#pragma HLS INTERFACE axis port=video_out bundle=OUTPUT_STREAM

#pragma HLS INTERFACE s_axilite port=rows bundle=CONTROL_BUS offset=0x14
#pragma HLS INTERFACE s_axilite port=cols bundle=CONTROL_BUS offset=0x1C
#pragma HLS INTERFACE s_axilite port=return bundle=CONTROL_BUS

#pragma HLS INTERFACE ap_stable port=rows
#pragma HLS INTERFACE ap_stable port=cols

    YUV_IMAGE img_0(rows, cols);
    YUV_IMAGE img_1(rows, cols);
#pragma HLS dataflow
    hls::AXIvideo2Mat(video_in, img_0);
    hls::GaussianBlur<3,3>(img_0,img_1,0,0);
    hls::Mat2AXIvideo(img_1, video_out);
}
```

Figure 68 HLS Function to perform the Gaussian Blur

Having written the code we wish to synthesise and implement in the Zynq SoC, the next thing we need to do is create a test bench so that we can check the functionality using both C and Co-Simulation before we include the core within our Vivado design.

With the core written, we want to be able to test it, initially using a C simulation to prove that it does what we desire and then again using co-simulation against the synthesised HDL to verify that the HDL functions as required. We can use the same test bench for both the C and HDL testing using the HLS environment. Within the test bench, we need to perform the following steps:

- Open the image we wish to work with into a cv::mat
- Convert the image from RGB to YUV
- Convert the cv::mat to IplImage
- Convert the IplImage into an AXI Stream
- Call the synthesizable function
- Convert the AXI Stream to an IplImage
- Convert the YUV image to RGB
- Write the image to file

This approach allows us to take in an image file, apply a Gaussian filter to it, and then save it to a file for later examination. The reason we do the conversion from RGB to YUV is because it is common to process images in this color space and the example we will eventually build from this preparatory work using the EVK employs this color space, so the input image must also be in this color space.

The color space we will be using is YUV 4:2:2, which means that the pixel value can be represented in 2 bytes. If we were to use the OpenCV cvt_color function, it would return a YUV 4:4:4 image. Therefore we need a specific routine to subsample the image, converting it from 4:4:4 to 4:2:2.

```
int main (int argc, char** argv) {
    Mat src_rgb = imread(INPUT_IMAGE);
    if (!src_rgb.data) {
        printf("ERROR: could not open or find the input image!\n");
        return -1;
    }
    Mat src_yuv(src_rgb.rows, src_rgb.cols, CV_8UC2);
    Mat dst_yuv(src_rgb.rows, src_rgb.cols, CV_8UC2);
    Mat dst_rgb(src_rgb.rows, src_rgb.cols, CV_8UC3);

    cvtcolor_rgb2yuv422(src_rgb, src_yuv);

    IplImage src = src_yuv;
    IplImage dst = dst_yuv;
    hls_image_filter(&src, &dst);
    cvtColor(dst_yuv, dst_rgb, CV_YUV2BGR_YUYV);
    imwrite(OUTPUT_IMAGE, dst_rgb);

}

void hls_image_filter(IplImage *src, IplImage *dst) {
    AXI_STREAM src_axi, dst_axi;
    IplImage2AXIvideo(src, src_axi);
    image_filter(src_axi, dst_axi, src->height, src->width);
    AXIvideo2IplImage(dst_axi, dst);
}
```

Figure 69 Test bench for the image processing core

Once we have generated both the function and the test bench file, the next step is to use C simulation to ensure that the function performs as desired. If it does not, we can quickly and easily modify the test bench

Image Processing With Xilinx Devices

or the source code and re simulate until we obtain the desired results.

Once assured that we've properly defined the desired function at the C level, we then synthesise the function using Vivado HLS to produce HDL. As part of the synthesis process, Vivado HLS estimates the resource utilization. For this example, where we are targeting the Zynq Z-7020 SoC, the estimate showed that the following resources would be required:

Utilization Estimates

□ Summary

Name	BRAM_18K	DSP48E	FF	LUT
DSP	-	-	-	-
Expression	-	-	0	1
FIFO	0	-	20	80
Instance	6	8	925	1463
Memory	-	-	-	-
Multiplexer	-	-	-	-
Register	-	-	-	-
Total	6	8	945	1544
Available	280	220	106400	53200
Utilization (%)	2	3	~0	2

Figure 70 Resource Estimation

We now wish to perform co-simulation with the synthesized output. This step allows us to use the C test bench to stimulate the generated HDL using an HDL simulation tool. The images below show the original input image and the resultant output image.

Figure 71 Original Image

Image Processing With Xilinx Devices

Figure 72 Co-Simulation Gaussian Blur 3x3

Once the co-simulation is complete, we can examine the resultant output image and, if we wish, the simulation waveforms. There is also a status report on the co-simulation, which not only reports the pass/fail status but also the function latency. We can compare this latency from co-simulation against the synthesis report, which also contains the expected latency. However, the expected latency shown in the synthesis report is based on handling the maximum row and column sizes while the latency estimation in the simulation results are based on the actual image size passed to it.

Cosimulation Report for 'image_filter'

Result

RTL	Status	Latency min	Latency avg	Latency max	Interval min	Interval avg	Interval max
VHDL	Pass	267803	267803	267803	0	0	0
Verilog	NA	NA	NA	NA	NA	NA	NA

Figure 73 Co Simulation Report

Once we are happy with the co-simulation and know that the function works as intended, the final step is to export the module into our Vivado IP library and insert the new IP module into our image-processing chain. This we can achieve very easily by using the Export RTL option and completing the configuration options as you desire as shown below:

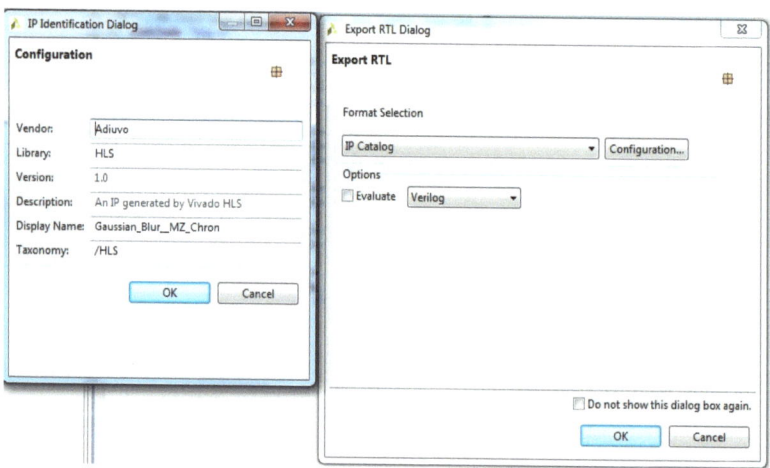

Figure 74 Packaging and Exporting Generated cores

This will package the IP. You will find the packaged IP core and a Zip file of the IP core within your project's solutions directory.

We can then open our Vivado design and import the IP Core we just created from the IP Catalog. However, to do this we first need to create an IP Repository within our project using the projects settings dialog on the IP tab:

Image Processing With Xilinx Devices

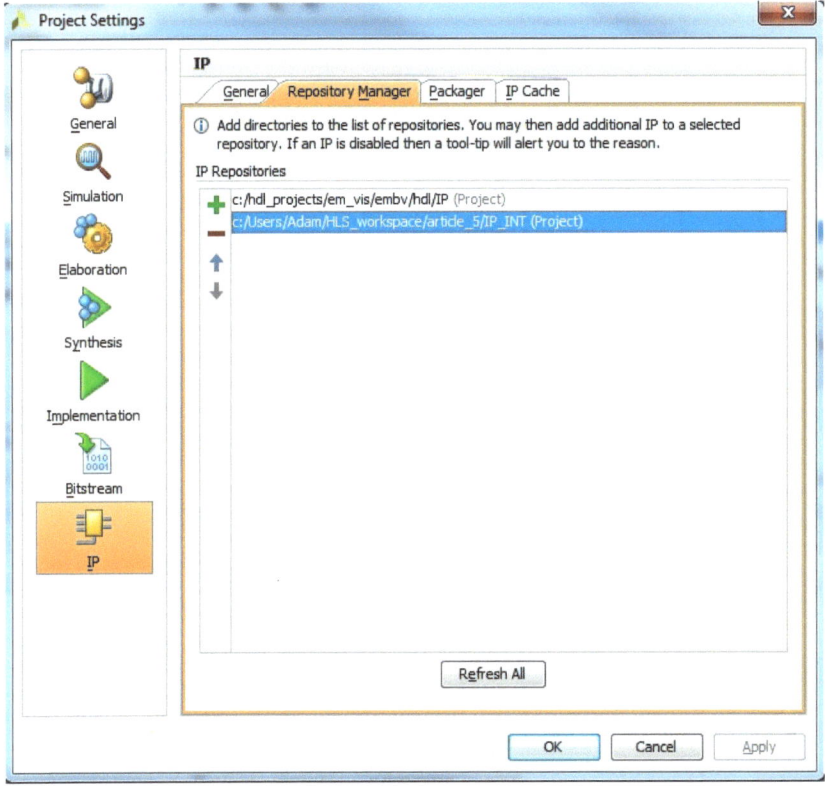

Figure 75 Creation of the IP repository with this example highlighted

After we create the IP Repository, it will not contain any IP Cores. We need to add cores to it using the IP Catalog. Within the IP Catalog, you should see the Repository that we just created. Right-click on the Repository and then select Add IP to Repository option.

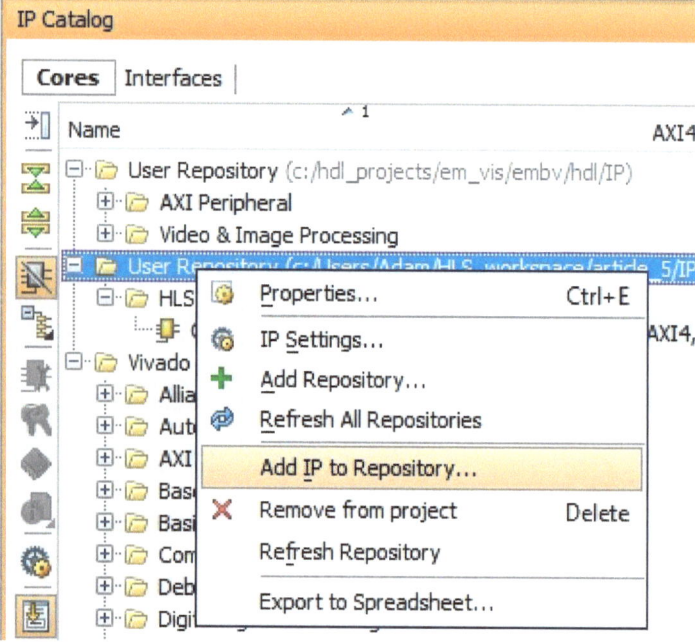

Figure 76 Adding Generated IP to the repository

This will open a dialog box so that we can select the IP Core we wish to add to the repository. We can select either the component.xml or the zipped archive. When this is complete, you will see the IP core located within the Repository, ready for use in a block diagram.

Figure 77 IP within the library to be added in to the design

Having now shown how we can quickly and easily get image processing functions up and running,

Image Processing With Xilinx Devices

TRANSFERING IMAGES FROM THE PL TO PS

One of the many use cases for the Zynq MPSoC is embedded vision, the APU, Mali GPU, Display Port and Programmable Logic mean it can address exciting applications like ADAS and Vision Guided Robotics with relative ease. When it comes to interfacing the image sensors with the Zynq MPSoC we can use the programmable logic to interface with a range of standards from MIPI, LVDS, Parallel, VoSPI etc.

Once the image is within the PL we can implement a image processing pipeline using existing IP cores from the Xilinx library or develop our own custom IP cores using High Level Synthesis. However, for many applications we need to be able to move the images into the PS domain before we can apply the exciting application level algorithms such as decision making or use with the reVISION acceleration stack.

Figure 78 The original MicroZed Evaluation kit & UltraZed board used for this demo

As such I thought I would kick off the fourth year of this blog with a look at how we can use VDMA in a Zynq MPSoC PL to transfer images from the PL to the PS DDR Memory.

To do this we will use the following IP blocks

- Zynq MPSoC core – Configured to enable both a Full Power Domain (FPD) AXI HP Master and FPD HPC AXI Slave, along with providing at least one PL clock and reset to the fabric.
- VDMA – Configured for write only operations, No FSync option & Genlock Mode of master
- Test Pattern Generator (TPG) – Configurable over the AXI Lite interface
- AXI Interconnects – Implements the Master and Slave AXI networks

In this example, the Test Pattern Generator once configured over AXI Lite, outputs test patterns which are then transferred into the processing system DDR memory. To demonstrate this has been successful we can examine the memory location using SDK.

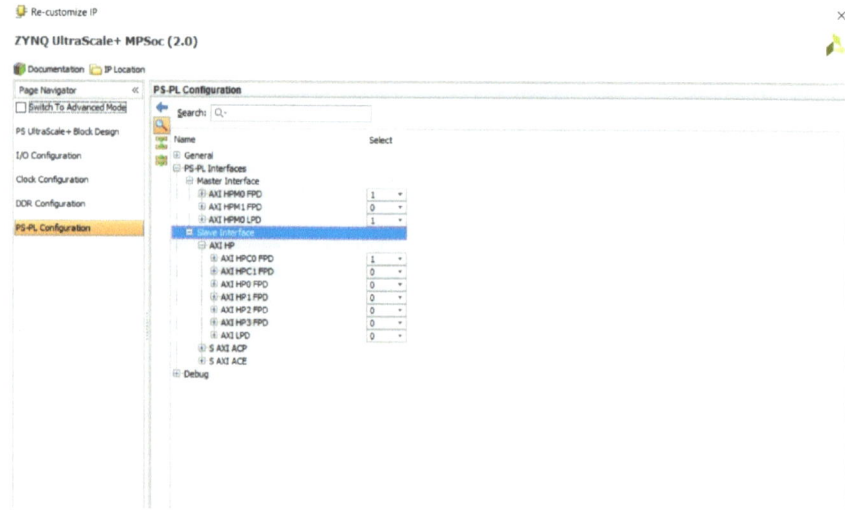

Figure 79 Enabling the FPD Master and Slave Interfaces

For this simple example both the AXI networks will be clocked at the

Image Processing With Xilinx Devices

same frequency being driven by PL_CLK_0 at 100 MHz.

For a deployed system, where an image sensor replaces the TPG as the image source we need to ensure the VDMA input channels clocks (Slave to Memory Map and Memory Map to Slave) are fast enough to support the pixel and frame rate required. For example a sensor resolution of 1280 pixels by 1024 lines at 60 frames per second requires a clock rate of at least 108 MHZ.

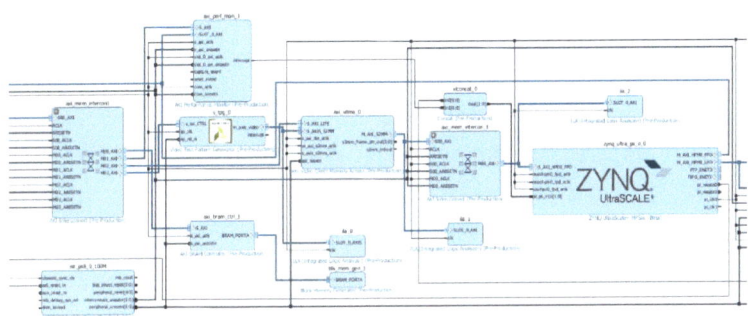

Figure 80 Block Diagram of the completed design

To aid visibility within this example I have included three ILA modules connected to the outputs of the Test Pattern Generator, AXI VDMA and the Slave Memory Interconnect. This enables the use of hardware manager in Vivado to verify the software has correctly configured the TPG and the VDMA to transfer the images. With the Vivado design complete and built the application SW to configure the TPG and VDMA to generate images and move them from the PL to the PS is very straight forward. To aid in the generation of the application the following APIs available under the BSP lib source directory are used AXIVDMA, V_TPG & Video Common. While the software itself performs the following:

1. Initialise the Test Pattern Generator and the AXI VDMA for use in the SW application
2. Configure the TPG to generate a test pattern configured as below
    a. Set the Image width 1280, Image Height 1080

b. Set the colour space of YCRCB 4:2:2 format
c. Set the Test Pattern Generator back ground pattern
d. Enable the test pattern generator and set for auto reloading
3. Configure the VDMA to write data into the PS memory
    a. Set up the VDMA parameters using a variable of the type XAxiVdma_DmaSetup – remember the horizontal size and stride are measure in bytes not pixels.
    b. Configure the VDMA with the setting defined above
    c. Set the VDMA frame store location address in the PS DDR
    d. Start VDMA transfer

The application will then start generation of test frames from the TPG into the PS DDR Memory, to ensure the DDR memory is updated for this example I disabled the caches. Examining the ILAs you will see the TPG generating frames and the VDMA transferring the stream into memory mapped format.

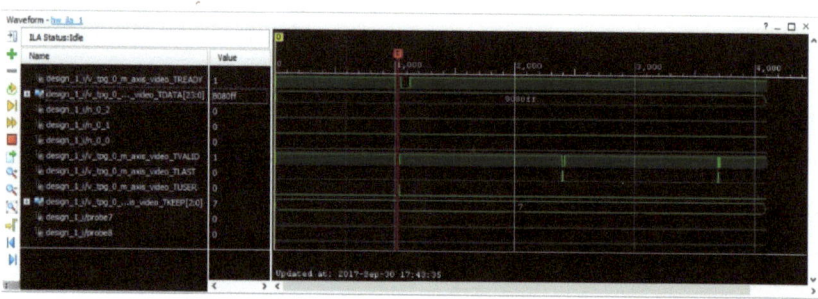

Figure 81 TPG output, TUSER indicates start of frame while TLAST indicates end of line

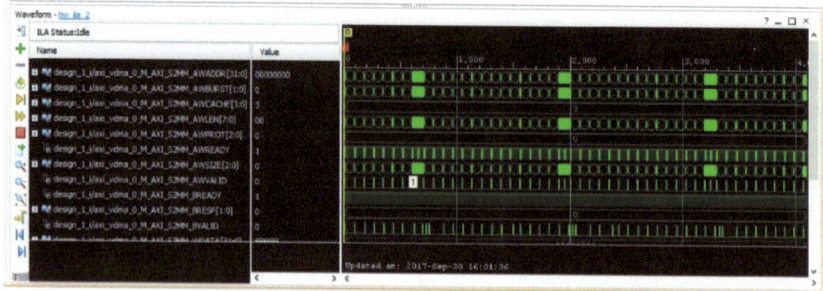

Figure 82 VDMA Memory Mapped Output to the PS.

Image Processing With Xilinx Devices

Examining the frame store memory location within the PS DDR using SDK demonstrates the pixel values are present.

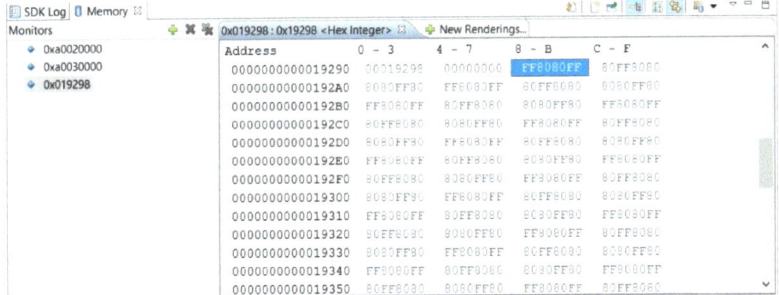

Figure 83 Test Pattern Pixel Values within the PS DDR Memory

If you are using a Zynq 7000 and not a Zynq MPSoC you can use the same approach in Vivado and SW by enabling the AXI GP master for the AXI Lite bus and AXI HP slave for the VDMA channel.

ADDRESSING VDMA ISSUES

Video Direct Memory Access (VDMA) is one of the key IP blocks used within many image-processing applications. It allows frames to be moved between the Zynq SoC's and Zynq UltraScale+ MPSoC's PS and PL with ease. Once the frame is within the PS domain, we have several processing options available. We can implement high-level image processing algorithms using open-source libraries such as OpenCV and acceleration stacks such as the Xilinx reVISION stack if we wish to process images at the edge. Alternatively, we can transmit frames over Gigabit Ethernet, USB3, PCIe, etc. for offline storage or later analysis.

It can be infuriating when our VDMA-based image-processing chain does not work as intended. Therefore, we are going to look at a simple VDMA example and the steps we can take to ensure that it works as desired.

The simple VDMA example shown below contains the basic elements needed to provide VDMA output to a display. The processing chain starts with a VDMA read that obtains the current frame from DDR memory. To correctly size the data stream width, we use an AXIS subset convertor to convert 32-bit data read from DDR memory into a 24-bit format that represents each RGB pixel with 8 bits. Finally, we output the image with an AXIS-to-video output block that converts the AXIS stream to parallel video with video data and sync signals, using timing provided by the Video Timing Controller (VTC). We can use this parallel video output to drive a VGA, HDMI, or other video display output with an appropriate PHY.

This example outlines a read case from the PS to the PL and corresponding output. This is a more complicated case than performing a frame capture and VDMA write because we need to synchronize video timing to generate an output.

Image Processing With Xilinx Devices

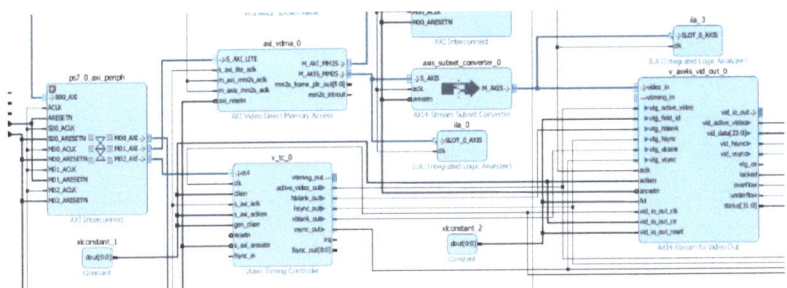

Figure 84 Simple VDMA-Based Image-Processing Pipeline

So what steps can we take if the VDMA-based image pipeline does not function as intended? To correct the issue:

1. **Check Reset and Clocks** as we would when debugging any application. Ensure that the reset polarity is correct for each module as there will be mixed polarities. Ensure that the pixel clock is correct for the required video timing and that it is supplied to both the VTC and the AXIS-to-Video Out blocks. While the clock required for the AXIS network must be able to support the image throughput.
2. **Check the Clock Enables** on both the VTC and AXIS to Video Out blocks are tied to the correct level to enable the clocks.
3. **Check that the VTC** is correctly configured, especially if you are using the AXI interface to define the configuration through the application software. When configuring the VTC using AXI, it is important to make sure we have set the source registers to the VTC generator, enabled register updates, and defined the timing parameters required.
4. **Check the connections** between the VTC and AXIS-to-Video-Out Blocks. Ensure that the horizontal and vertical blanking signals are also connected along with the horizontal and vertical syncs.
5. **Check the AXIS-to-Video-Out** configuration. If we are using VDMA, the timing mode of the AXIS-to-Video-Out block should be set to master. This enables the AXIS-to-Video-Out block to assert back pressure on the AXIS data stream to halt the frame buffer output. This mechanism permits the AXIS-to-Video-Out

block to manage the flow of pixels by enabling synchronization and lock. You may also want to increase the size of the internal buffer from the default.

6. **Check that the AXIS-to-Video-Out** VTC_ce signal is not connected to the VTC gen clock enable as is the case when configured for slave operation. This will prevent the AXIS-to-Video-Out block from being able to lock to the AXIS video stream.
7. **Insert ILA's.** Inserting these within the design allow us to observe the detailed workings of the AXI buses. When commissioning a new image processing pipeline, I insert ILA blocks on the VTC output and the VDMA MM-to-AXIS port so that I can observe the generated timing signals and VDMA output stream. When observing the AXI Stream the tuser signal identifies the start of frame and the tlast signal represents the end of line. You may also want to observe the AXIS-to-Video-Out 32-bit status output, which provides indication of the locked status along with additional debug information.
8. **Ensure that HSize and Stride** are set correctly. These are defined by the application software and configure the VMDA with frame-store information. HSize represents the horizontal size of the image and Stride represents the distance in memory between the image lines. Both HSize and Stride are defined in bytes. As such, when working with U32 or U16 types, take care to correctly set these values to reflect the number of bytes used.

Hopefully by the time you have checked these points, the issue with your VDMA based image processing pipeline will have been identified and you can start developing the higher-level image processing algorithms needed for the application.

Image Processing With Xilinx Devices

VERIFYING VIDEO IMAGES IN ZYNQ SOC AND ZYNQ ULTRASCALE+ MPSOC MEMORY

Getting the best performance from our embedded-vision systems often requires that we can capture frames individually for later analysis in addition to displaying them. Programs such as Octave, Matlab, or Image J can analyse these captured frames, allowing us to examine parameters such as:

- Compare the received pixel values against those expected for a test or calibration pattern.
- Examine the Image Histogram, enabling histogram equalization to be implemented if necessary.
- Ensure that the integration time of the imager is set correctly for the scene type.
- Examine the quality of the image sensor to identify defective pixels—for example dead or stuck-at pixels.
- Determine the noise present in the image. The noise present will be due to both inherent imager noise sources—for example fixed pattern noise, device noise and dark current—and also due to system noise as coupled in via power supplies and other sources of electrical noise in the system design.

Typically, this testing may occur in the lab as part of the hardware design validation and is often performed before the higher levels of the application software are available. Such testing is often implemented using a bare-metal approach on the processor system.

If we are using VDMA, the logical point to extract the captured data is from the frame buffer in the DDR SDRAM attached to the Zynq SoC's or MPSoC's PS. There are two methods we can use to examine the contents of this buffer:

- Use XSCT terminal to read out the frame buffer and post process it using a TCL script.

- Output the frame buffer over RS232 or Ethernet using the Light Weight IP Stack and then capturing the image data in a terminal for post processing using a TCL file.

For this example, I am going to use the UltraZed design we created a few weeks ago to examine PL-to-PS image transfers in the Zynq UltraScale+ MPSoC (see here). This design rather helpfully uses the test pattern generator to transfer a test image to a frame buffer in the PS-attached DDR SDRAM. In this example, we will extract the test pattern and convert it into a bit-map (BMP) file. Once we have the bit-map file, we can read it into the analysis program of choice.

BMP files are very simple. In the most basic format, they consist of a BMP Header, Device Independent Bitmap (DIB) Header, and the pixel array. In this example the pixel array will consist of 24-bit pixels, using eight bits each for blue, green and red pixel values.

It is important to remember two key facts when generating the pixel array. First, when generating the pixel array each line must be padded with zeros so that its length is a multiple of four, allowing for 32-bit word access. Second, the BMP image is stored upside down in the array. That is the first line of the pixel array is the bottom line of the image.

Combined, both headers equal 54 bytes in length and are structured as shown below:

Address	Size (Bytes)	Function	Value
00	2	Identifier	0x42 0x4D
02	4	Bit Map Size	Size in Bytes
06	2	Reserved	
08	2	Reserved	
0A	4	Offset to Pixel Array	0x36 for this example

Figure 85 Bitmap Header Construction

Image Processing With Xilinx Devices

Address	Size (Bytes)	Function	Value
0E	4	DIB Header Size	0x28
12	4	Bit Map Width	0x258 for this example
16	4	Bit Map Height	0x1E0 for this example
1A	2	Colour Planes	Must be 0x01
1C	2	Bits per Pixel	0x18 for this example
1E	4	Compression Format	0 for this example
22	4	Image Size	0 for this example
26	4	Horizontal Resolution (pixels per meter)	0xB13
2A	4	Vertical Resolution (pixels per meter)	0xB13
2E	4	Number of colours in palette	0 for this example
32	4	Number of important colours	0 for this example

Figure 86 DIB Header Construction

Having understood what is involved in creating the file, all we need to do now is gather the pixel data from the PS-attached DDR SDRAM and output it in the correct format.

As we have done several times before in this blog, when we extract the pixel values it is a good idea to double check that the frame buffer contains pixel values. We can examine the contents of the frame buffer using the memory viewer in SDK. However, the view we choose will ease our understanding of the pixel values and hence the frame. This is due to how the VDMA packs the pixels into the frame buffer.

The default view for the Memory viewer is to display 32-bit words as shown below:

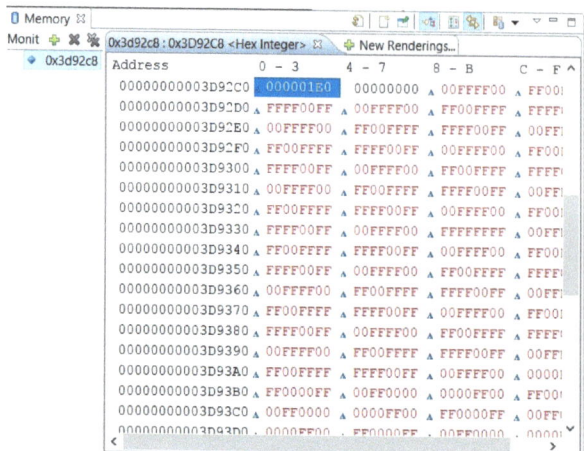

Figure 87 TPG Test Pattern in memory

The data we are working with has a pixel width of 24 bits. To ensure efficient use of the DDR SDRAM memory, the VDMA packs the 24-bit pixels into 32-bit values, splitting pixels across locations. This can make things a little confusing when we look at the memory contents for expected pixel values. Because we know the image is formatted as 8-bit RGB, a better view is to configure the memory display to list the memory contents in byte order. We then know that each group of three bytes represents one pixel.

Image Processing With Xilinx Devices

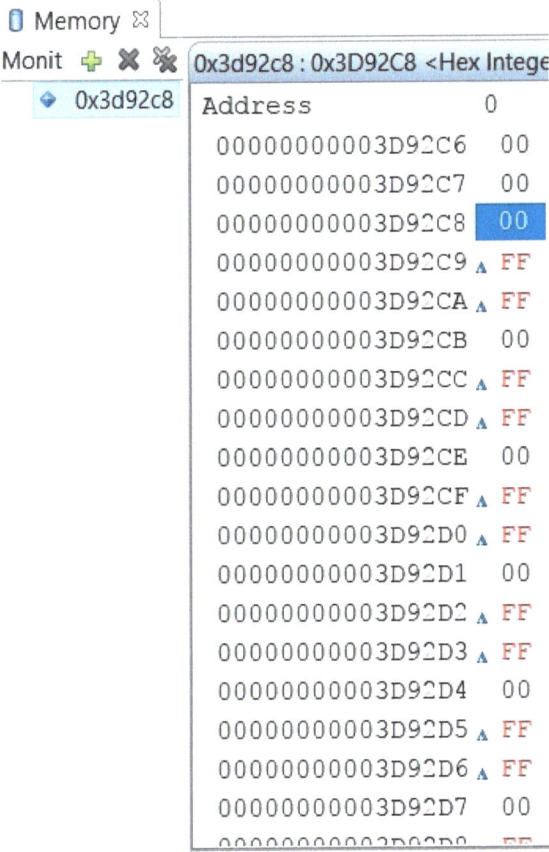

Figure 88 TPG Test Pattern in memory Byte View

Having confirmed that the frame buffer contains image data, I am going to output the BMP information over the RS232 port for this example. I have selected this interface because it is the simplest interface available on many development boards and it takes only a few seconds to read out even a large image.

The first thing I did in my SDK application was to create a structure that defines the header and sets the values as required for this example:

```
Header.Type = 0x424d;
Header.Size = 0xA61E0D00;
Header.Reserve1 = 0x0000;
Header.Reserve2 = 0x0000;
Header.OffBits = 0x36000000;
Header.biSize = 0x28000000 ;
Header.biWidth = 0x58020000;
Header.biHeight = 0xE0010000;
Header.biPlanes = 0x0100;
Header.biBitCount = 0x1800;
Header.biCompression =0x00000000;
Header.biSizeImage = 0x00000000;
Header.biXPelsPerMeter = 0x130B0000;
Header.biYPelsPerMeter = 0x130B0000;
Header.biClrUsed = 0x00000000;
Header.biClrImportant = 0x00000000;
```

Figure 89 Header Structure in the application

I then created a simple loop that creates three u8 arrays, each the size of the image. There is one array for each color element. I then used these arrays with the header information to output the BMP information, taking care to use the correct format for the pixel array. A BMP pixel array organizes the pixel element as Blue-Green-Red:

Image Processing With Xilinx Devices

```
addr = vdmaDMA.FrameStoreStartAddr[0];
while(1){
 for (y=0;y<480;y++){
    for (i = 0; i< 600;i++){
        red[y][i]= Xil_In8(addr);
        blue[y][i]= Xil_In8(addr+1);
        green[y][i]= Xil_In8(addr+2);
        addr = addr +3;
    }
    addr = addr + 600;
 }
 y =0;
 i =0;

 printf("%04x%08x%04x%04x%08x%08x%08x%08x%04x%04x%08x%08x%08x%08x%08x%08x",
        Header.Type,Header.Size,Header.Reserve1,Header.Reserve2,
        Header.OffBits,Header.biSize,Header.biWidth,
        Header.biHeight,Header.biPlanes,Header.biBitCount,
        Header.biCompression,Header.biSizeImage,
        Header.biXPelsPerMeter,Header.biYPelsPerMeter,
        Header.biClrUsed,Header.biClrImportant);

 for (y=0;y<480;y++){
    for (i = 0; i< 600;i++){

        printf("%02x%02x%02x",blue[y][i],green[y][i],red[y][i]);

    }
    //printf("\n");
 }
break;
```

Figure 90 Body of the Code to Output the Image

Wanting to keep the processes automated and without the need to copy and paste to capture the output, I used Putty as the terminal program to receive the output data. I selected Putty because it is capable of saving received data a log file.

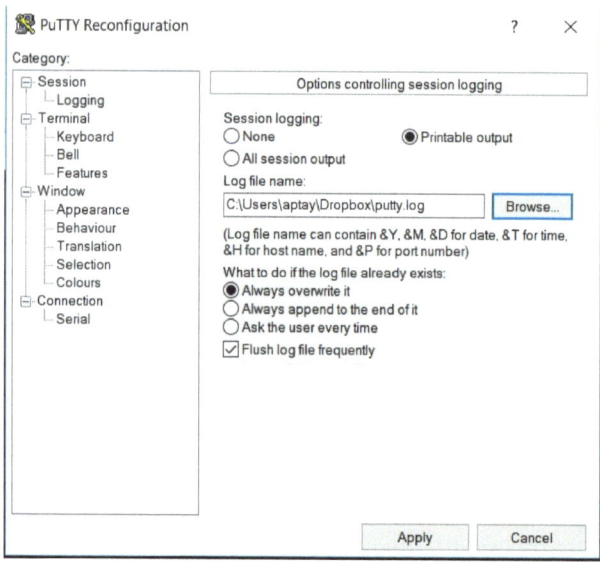

Figure 91 Putty Configuration for logging

Of course, this log file contains an ASCII representation of the BMP. To view it, we need to convert it to a binary file of the same values. I wrote a simple TCL script to do this. The script performs the conversion, reading in the ASCII file and writing out the binary BMP File.

Image Processing With Xilinx Devices

Figure 92 TCL ASCII to Binary Conversion Widget

With this complete, we have the BMP image which we can load into Octave, Matlab, or another tool for analysis. Below is an example of the tartan color-bar test pattern that I captured from the Zynq frame buffer using this method:

Figure 93 Generated BMP captured from the PS DDR

Now if we can read from the frame buffer, then it springs to mind that we can use the same process to write a BMP image into the frame buffer. This can be especially useful when we want to generate overlays and use them with the video mixer.

Image Processing With Xilinx Devices

VIDEO MIXING

So far, all the image processing examples, have used only one sensor and hence one video stream within the Zynq or Zynq MPSoC PL. However, if we want to work with multiple sensors or overlay information like telemetry etc on a frame we need to do some video mixing.

Video mixing enables several different video streams to be merged together, to create one output video stream. In our designs we can use this in several ways

1. Tile together multiple video streams to be displayed on a larger display for example stitching multiple images into a 4K display.
2. Blend together multiple image streams as vertical layers to create one final image, for example adding an overlay or performing sensor fusion

To do this within our Zynq or Zynq MPSoC system we use the [Video Mixer IP core](), which comes with the Vivado IP library. This core mixes together up to 8 image streams, plus a final logo layer. The image streams are provided to the core via AXI Streaming or AXI memory mapped inputs. With the mixed video frame output using a AXI Stream such that we can implemented it within the image processing path. To give a demonstration of the how we can use the video mixer, I am going to update the MiniZed FLIR Lepton project to use the 10-inch touch display and merge a second video stream using a test pattern generator. Using the 10-inch touch display gives me a larger screen to demonstrate the concept and it has been sat in my office for a while now. Upgrading to use the 10-inch display is easy, all we need to do in the Vivado design is increase the pixel clock frequency (fabric clock 2) from 33.33MHz to 71.1MHz. Along with adjusting the clock frequency setting in the ALI3 controller block to 71.1MHz.

Within the MiniZed Vivado design include a video mixer, enabling layer one and select a streaming interface with global alpha control enabled.

Enabling global alpha control enables for a layer allows the video mixer to blend the alpha on a pixel by pixel basis. This allows pixels to be merged according to the defined alpha value rather than just over riding the pixel on the layer beneath. The alpha value for each layer ranges between 0 transparent and 1 opaque, each layer's alpha value is defined within an 8-bit register.

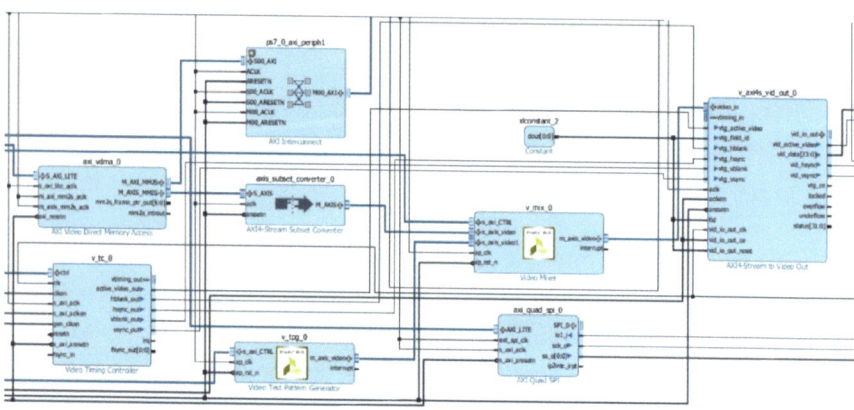

Figure 94 Insertion of the Video Mixer and Video Test Pattern Generator

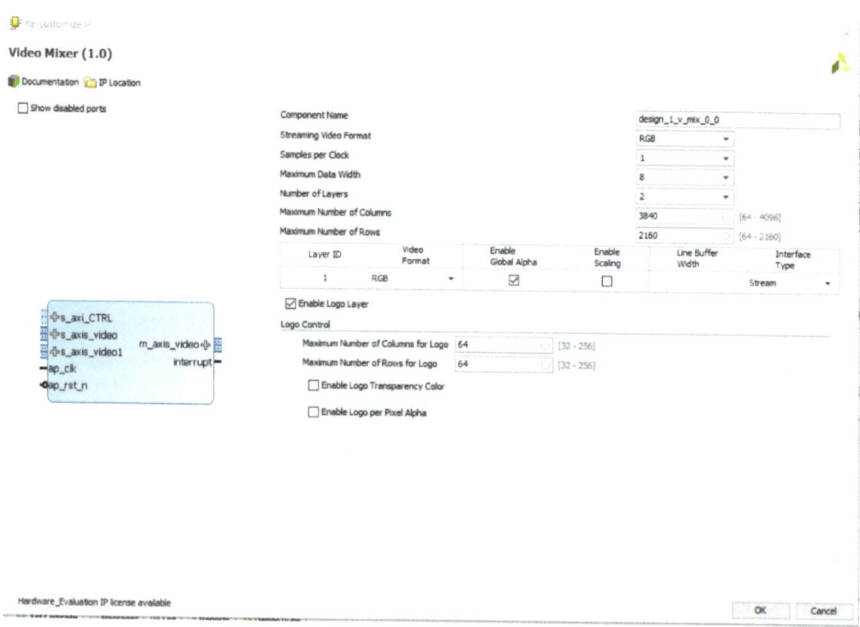

Figure 95 Enabling layer 1, for AXI streaming and Global Alpha Blending

Image Processing With Xilinx Devices

The FLIR images proves the first image, however, we need a second image stream instantiate a video test pattern generator core and connect its output to the video mixer layer 1 input. For the video mixer and test pattern generator, be sure to use the high-speed video clock used in the image processing chain. Build the design and export it to SDK.

To configure the video mixer in SDK we use the API xv_mix.h this provides the functions needed to control the video mixer.

The principle of the mixer is simple there is a master layer the vertical and horizontal size of which is declared using the API. For this example, using the 10-inch display we set this to 1280 pixels by 800 lines. We can then fill this image space using the layers, either tiling or overlapping them as desired for our application.

To do this along with each layer having an alpha register to control blending there is also X and Y origin registers along with a height and width registers, these enable the mixer to create the final image. Positional location for a layer which does not fill the entire display area is referenced from the top left of the display.

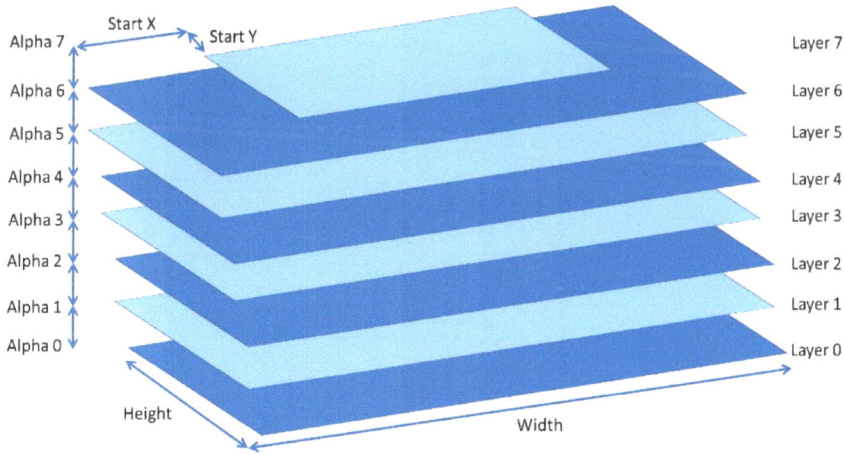

Figure 96 Video Mixing Layers, concept layer 7 is a reduced size image.

To demonstrate, this in action the test pattern generator was used to create a 200 by 200 checkerboard pattern. With the video mixer configured first with the TPG layer alpha set to opaque such that it would override the FLIR Image. Then again with the alpha set to a lower value enabling merging of the two layers

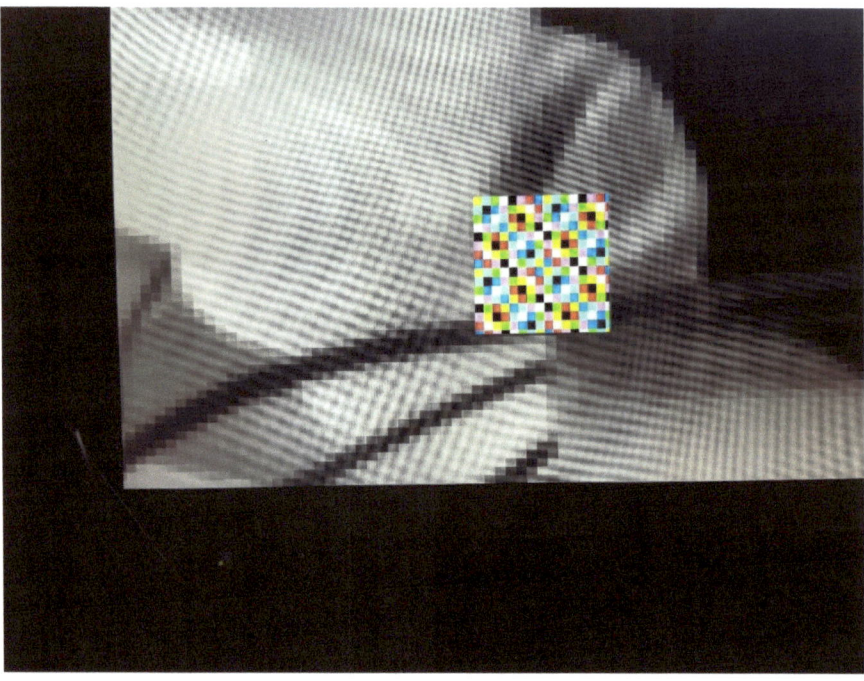

Figure 97 Test Pattern FLIR & Test Pattern Generator Layers merged, test pattern has higher alpha.

Image Processing With Xilinx Devices

Figure 98 Test Pattern FLIR & Generator Layers merged, test pattern alpha lower.

We can also use the video mixer to tile images as shown below, to create this image I added in three more test pattern generators.

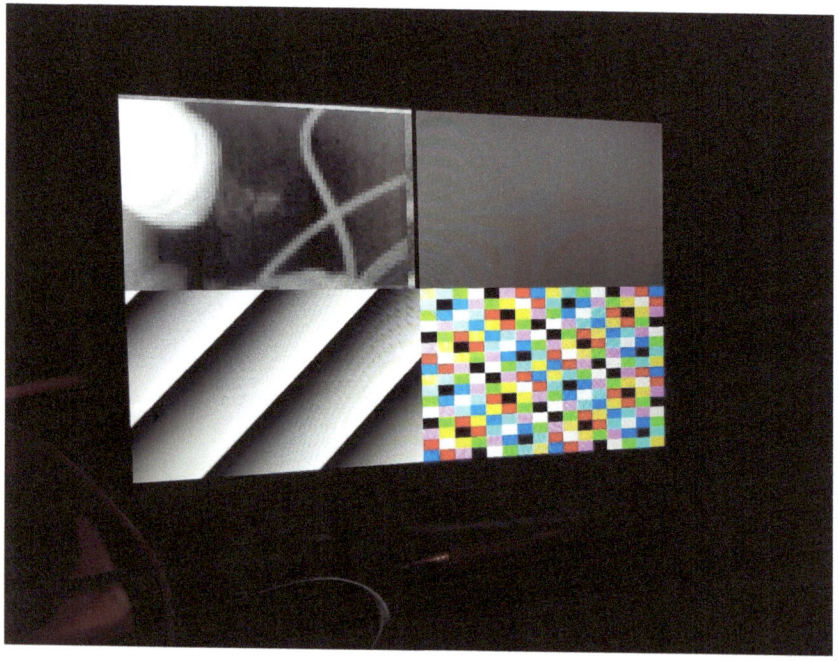

Figure 99 Four video streams tiled using the mixer

The video mixer is a good tool in our tool box when creating image processing or display solutions. It is very useful if we want to merge together the outputs of cameras working in different elements of the electromagnetic spectrum which we will examine in the future.

Image Processing With Xilinx Devices

SYNCHRONISING OUTPUT VIDEO

Recently, I have been doing some significant development work with the Avnet Embedded Vision Kit (EVK) significantly (for more info on the EVK and its uses see Issues 114 to 126 of the MicroZed Chronicles). As part my development, I wanted to synchronize the EVK display output with an external source—also useful if we desire to synchronize multiple image streams.

Implementing this is straight forward provided we have the correct architecture. The main element we need is a buffer between the upstream camera/image sensor chain and the downstream output-timing and -processing chain. VDMA (Video Direct Memory Access) provides this buffer by allowing us to store frames from the upstream image-processing pipeline in DDR SDRAM and then reading out the frames into a downstream processing pipeline with different timing.

The architectural concept appears below:

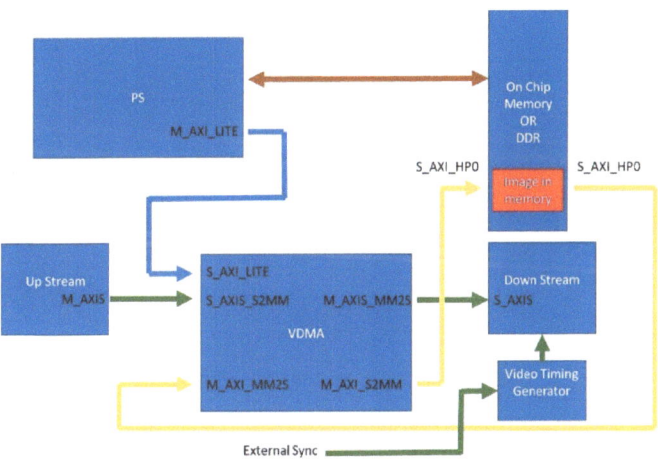

Figure 100 VDMA buffering between upstream and downstream with external sync

For most downstream chains, we use a combination of the video timing controller (VTC) and AXI Stream to Video Out IP blocks, both provided in the Vivado IP library. These two IP blocks work together. The VTC provides output timing and generates signals such as VSync and HSync. The AXI Stream to Video Out IP Block synchronizes its incoming AXIS stream with the timing signals provided by the VTC to generate the output video signals. Once the AXI Stream to Video Out block has synchronized with these signals, it is said to be locked and it will generate output video and timing signals that we can use.

The VTC itself is capable of both detecting input video timing and generating output video timing. These can be synchronized if you desire. If no video input timing signals are available to the VTC, then the input frame sync pulse (FSYNC_IN) serves to synchronize the output timing.

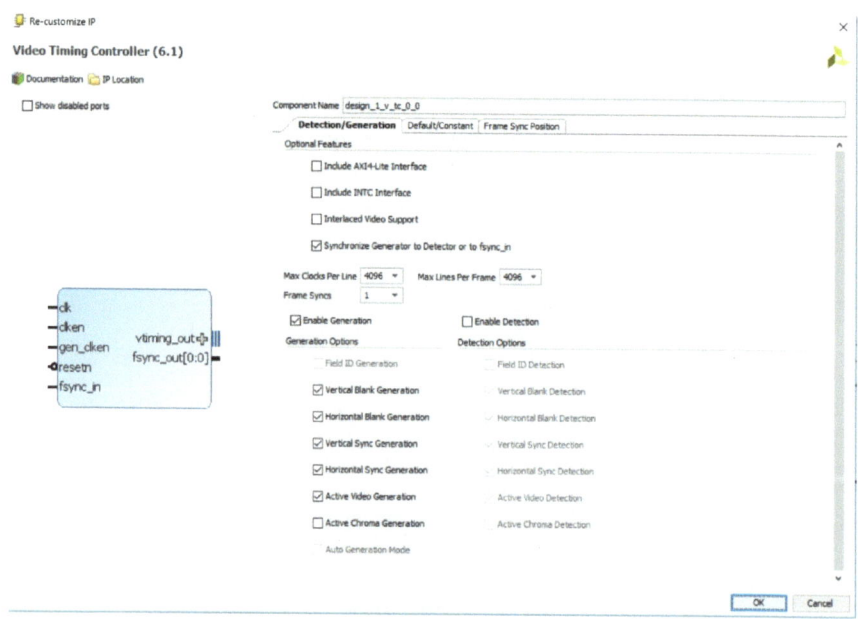

Figure 101 Enabling Synchronization with FSYNC_IN or the Detector

If FSYNC_IN alone is used to synchronize the output, we need to use not only FSYNC_IN but also the VTC-provided frame sync out (FSYNC_OUT) and GEN_CLKEN to ensure correct synchronization. GEN_CLKEN is an input enable that allows the VTC generator output stage to be clocked.

Image Processing With Xilinx Devices

The FSYNC_OUT pulse can be configured to occur at any point within the frame. For this application, is has been configured to be generated at the very end of the frame. This configuration can take place in the VTC re-configuration dialog within Vivado for a one-time approach or, if an AXI Lite interface is provided, it can be positioned using that during run time.

The algorithm used to synchronize the VTC to an external signal is:

1) Generate a 1-clock-wide pulse on FSYNC_IN reception
2) Enable GEN_CLK
3) Wait for the FSYNC_OUT to be received
4) Disable GEN_CLK
5) Repeat from step 1

Should GEN_CLK not be disabled, the VTC will continue to run freely and will generate the next frame sequence. Issuing another FSYNC_IP while this is occurring will not result in re-synchronisation but will result in the AXI Stream to Video Out IP block being unable to synchronize the AXIS video with the timing information and losing lock.

Therefore, to control the enabling of the GEN_CLKEN we need to create a simple RTL block that implements the algorithm above.

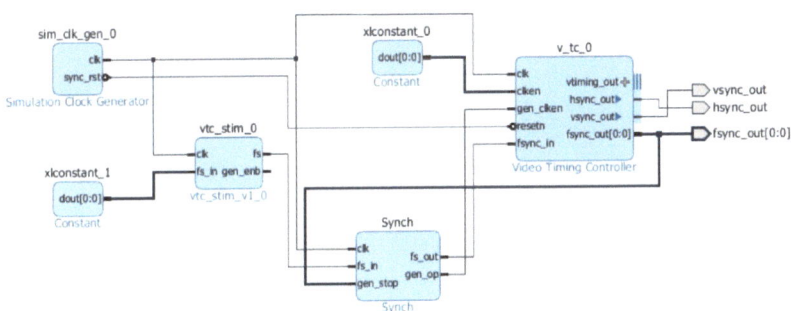

Figure 102 Vivado Project Demonstrating the concept

When simulated, this design resulted in the VTC synchronizing to the FSYNC_IN signal as intended. It also worked the same when I implemented it in my EVK kit, allowing me to synchronize the output to an external trigger.

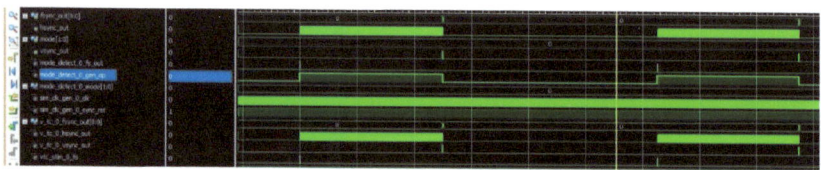

Figure 103 Simulation Results

Image Processing With Xilinx Devices

TIPS FOR BETTER IMAGE PROCESSING SYSTEMS

1. **Design in Flexibility from the Beginning**

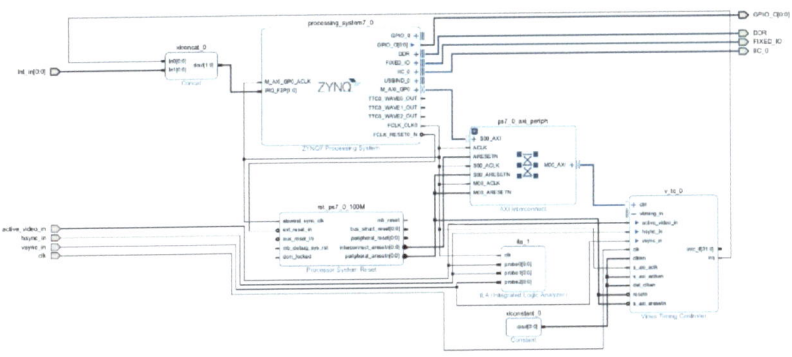

Figure 104 Video Timing Controller used to detect the incoming video standard

Use the flexibility provided by the Video Timing Controller (VTC) and reconfigurable clocking architectures such as Fabric Clocks, MMCM, and PLLs. Using the VTC and associated software running on the PS (processor system) in the Zynq SoC and Zynq UltraScale+ MPSoC, it is possible to detect different video standards from an input signal at run time and to configure the processing and output video timing accordingly. Upon detection of a new video standard, the software running on the PS can configure new clock frequencies for the pixel clock and the image-processing chain along with re-configuring VDMA frame buffers for the new image settings. You can use the VTC's timing detector and timing generator to define the new video timing. To update the output video timings for the new standard, the VTC can use the detected video settings to generate new output video timings.

2. **Convert input video to AXI Interconnect as soon as possible to leverage IP and HLS**

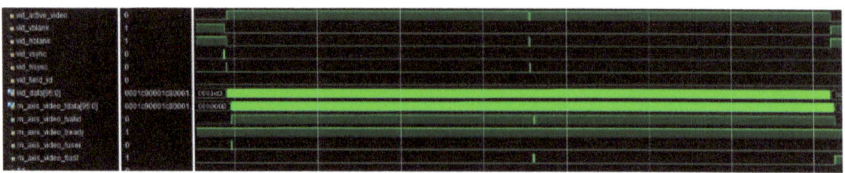

Figure 105 Converting Data into the AXI Streaming Format

Vivado provides a range of key IP cores that implement most of the functions required by an image processing chain—functions such as Color Filter Interpolation, Color Space Conversion, VDMA, and Video Mixing. Similarity Vivado HLS can generate IP cores that use the AXI interconnect to ease integration within Vivado designs. Therefore, to get maximum benefit from the available IP and tool chain capabilities, we need to convert our incoming video data into the AXI Streaming format as soon as possible in the image-processing chain. We can use the Video-In-to-AXI-Stream IP core as an aid here. This core converts video from a parallel format consisting of synchronization signals and pixel values into our desired AXI Streaming format. A good tip when using this IP core is that the sync inputs do not need to be timed as per a VGA standard; they are edge triggered. This eases integration with different video formats such as Camera Link, with its frame-valid, line-valid, and pixel information format, for example.

3. Use Logic Debugging Resources

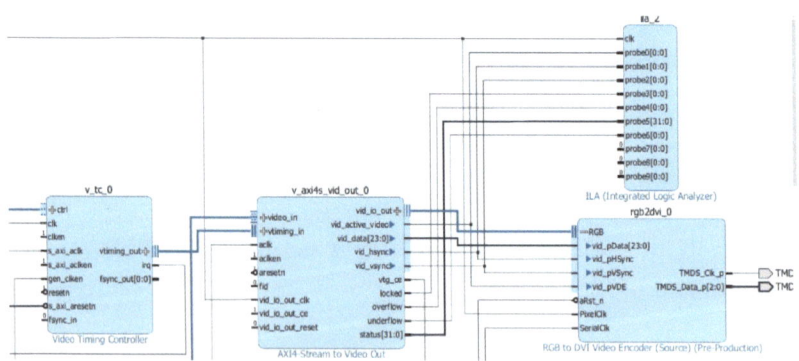

Figure 106 Insertion of the ILA monitoring the output stage

Image Processing With Xilinx Devices

Insert integrated logic analyzers (ILAs) at key locations within the image-processing chain. Including these ILAs from day one in the design can help speed commissioning of the design. When implementing an image-processing chain in a new design, I insert ILA's as a minimum in the following locations:

- Directly behind the receiving IP module—especially if it is a custom block. This ILA enables me to be sure that I am receiving data from the imager / camera.
- On the output of the first AXI Streaming IP Core. This ILA allows me to be sure the image-processing core has started to move data through the AXI interconnect. If you are using VDMA, remember you will not see activity on the interconnect until you have configured the VDMA via software.
- On the AXI-Streaming-to-Video-Out IP block, if used. I also consider connecting the video timing controller generator outputs to this ILA as well. This enables me to determine if the AXI-Stream-to-Video-Out block is correctly locked and the VTC is generating output timing.

When combined with the test patterns discussed below, insertion of ILAs allows us to zero in faster on any issues in the design which prevent the desired behavior.

4. **Select an Imager / Camera with a Test Pattern capability**

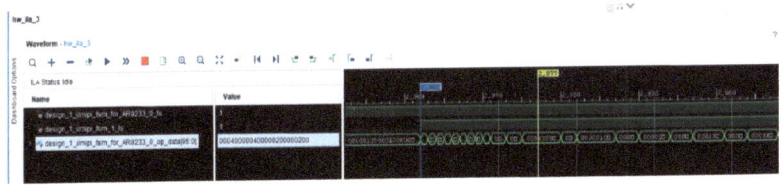

Figure 107 Incorrectly received incrementing test pattern captured by an ILA

If possible when selecting the imaging sensor or camera for a project, choose one that provides a test pattern video output. You can then use this standard test pattern to ensure the reception, decoding, and image-processing chain is configured correctly because you'll know exactly what the original video signal looks like. You can combine the imager/camera test pattern with ILAs connected close to the data reception module to determine if any issues you are experiencing when displaying an image is internal to the device and the image processing chain or are the result of the imager/camera configuration.

We can verify the deterministic pixel values of the test pattern using the ILA. If the pixel values, line length, and the number of lines are as we expect, then it is not an imager configuration issue. More likely you will find the issue(s) within the receiving module and the image-processing chain. This is especially important when using complex imagers/cameras that require several tens, or sometimes hundreds of configuration settings to be applied before an image is obtained.

5. **Include a Test Patter Generator in your Zynq SoC, Zynq UltraScale+ MPSoC, or FPGA design**

Figure 108 Tartan Color Bar Test Pattern

Image Processing With Xilinx Devices

If you include a test-pattern generator within the image-processing chain, you can use it to verify the VDMA frame buffers, output video timing, and decoding prior to the integration of the imager/camera. This reduces integration risks. To gain maximum benefit, the test-pattern generator should be configured with the same color space and resolution as the final imager. The test pattern generator should be included as close to the start of the image-processing chain as possible. This enables more of the image-processing pipeline to be verified, demonstrating that the image-processing pipeline is correct. When combined with test pattern capabilities on the imager, this enables faster identification of any problems.

6. **Understand how Video Direct Memory Access stores data in memory**

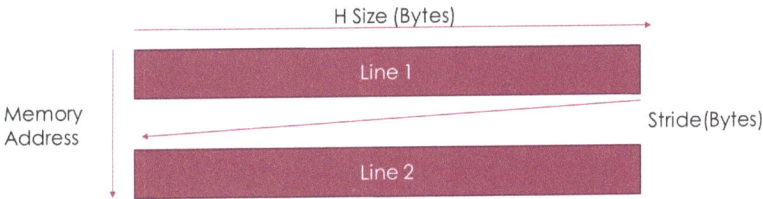

Video Direct Memory Access (VDMA) allows us to use the processor DDR memory as a frame buffer. This enables access to the images from the processor cores in the PS to perform higher-level algorithms if required. VDMA also provides the buffering required for frame-rate and resolution changes. Understanding how VDMA stores pixel data within the frame buffers is critical if the image-processing pipeline is to work as desired when configured.

One of the major points of confusion when implementing VDMA-based solutions centers around the definition of the frame size within memory. The frame buffer is defined in memory by three parameters: Horizontal Size (HSize), Vertical Size (VSize). and Stride. The two parameters that define the Horizontal Size of the image are the HSize and the stride of the image. Like VSize, which defines the number of lines in the image, the HSize defines the length of each line. However instead of being measured in pixels the horizontal size is measured in bytes. We therefore need to know how many bytes make up each pixel.

The Stride defines the distance between the start of one line and another. To gain efficient use of the DDR memory, the Stride should at least equal the horizontal size. Increasing the Stride introduces a gap between lines. Implementing this gap can be very useful when verifying that the imager data is received correctly because it provides a clear indication of when a line of the image starts and ends with memory.

These six simple techniques have helped me considerably when creating image processing examples for this blog or solutions for clients and they significantly ease both the creation and commissioning of designs.

Image Processing With Xilinx Devices

FLIR LEPTON EXAMPLE

The Lepton IR camera from FLIR is an 80x60-pixel (Lepton 2) or 160x120-pixel (Lepton 3) long-wave infra-red (LWIR) camera module. As a microbolometer-based thermal sensor, it operates without the need for cryogenic cooling, unlike HgCdTe-based sensors. Instead, a microbolometer works by each pixel changing resistance when IR radiation strikes it. This resistance change defines the temperatures in the scene. Typically, microbolometer-based thermal imagers have much-reduced resolution when compared to a cooled imager. They do however make thermal-imaging systems simpler to create.

Avnet's Zynq-based MiniZed is one of the most interesting dev boards we have looked at in this series. Thanks to its small form factor and its WiFi and Bluetooth capabilities, it is ideal for demonstrating Internet of Things (IoT) applications. We are now going to combine the FLIR Lepton camera module with the MiniZed and use them both to create a simple IOT application.

Figure 109 MiniZed with FLIR Lepton

The approach I am going to follow for this demonstration is to update the MiniZed PetaLinux hardware design to do the following:

- Interface with the FLIR Lepton camera module
- Implement a video-processing pipeline that supports a 7-inch touch display connected to the MiniZed's Pmod ports

The use of the local 7-inch touch display has two purposes. First, it demonstrates that the FLIR Lepton camera and the MiniZed are correctly working before I invest too much time in getting WiFi image transmission working. Second, the touch display could be used for local control and display, if required in an industrial (IIoT) application for example.

Opening the existing MiniZed Vivado project, you will notice it contains the Zynq (for the first time a single core Zynq) and an RTL block that interfaces with the WiFi and Bluetooth radio modules. This interface uses processing systems' (PS') SDIO0 for the WiFi interface and UART0 for Bluetooth. When we develop software, we must therefore remember to define the STDIN/STDOUT as being PS UART1 if we need a UART for debugging.

To this diagram we will add the following IP Blocks:

- Quad SPI Core – Configured for single-mode operation. Receives the VoSPI from the Lepton.
- Video Timing Controller – Generates the video timing signals for display output.
- VDMA – Reads an image from the PS DDR and converts it into a PL (programmable logic) AXI Stream.
- AXI Stream to Video Out – Converts the AXI Streamed video data to parallel video with timing synchronization provided by the Video Timing Core.
- Zed_ALI3_Controller – Display controller for the 7-inch touch-screen display.

Image Processing With Xilinx Devices

The Zed_ALI3_Controller IP block can be downloaded from the AVNET GitHub. Once downloaded, running the TCL script within the Vivado project will create an IP block we can include in our design.

The clocking architecture is now a little more complicated and includes the new Zed_ALI3_Controller block. This module generates the pixel clock, which is supplied to the VTC and the AXIS to Video blocks. Zynq-generated clocks provide the reference clock to the Zed_ALI3_Controller (33.33MHz) and the AXI Networks.

This demonstration uses two AXI networks. The first is the General-Purpose network. Te software uses this GP AXI network to configure IP blocks within the PL including the VDMA and VTC.

The second AXI network uses the High Performance AXI interface to transfer images from the PS DDR memory into the image-processing stream in the PL.

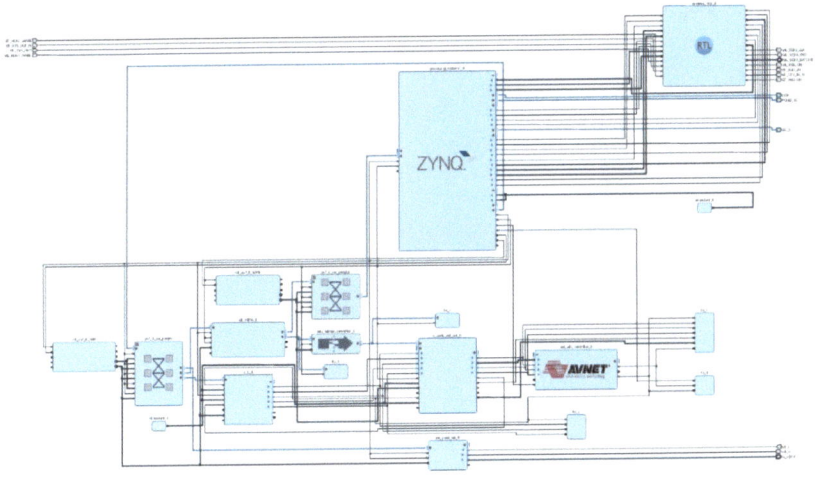

Figure 110 The complete block diagram.

To connect the FLIR Lepton camera module, we will connect to the MiniZed shield connector, making use of the shield's I2C and SPI connections.

The I2C pins are mapped into the constraints file already used for the temperature and motion sensors. Therefore, all we need to do is add the SPI I/O pin locations and standards.

The FLIR Lepton camera's AREF supply pin is not enabled. To power the camera on the shield connector as in the previous example, we take 5V power from a flying lead connected to the opposite shield connector's 5V supply and the back of the FLIR Lepton camera.

Figure 111 FLIR Lepton Connected to the MiniZed in the Shield Header

We'll need both Pmod connectors To output the image to the 7-inch display. The pin-out required appears below. The differential pins on the Pmod connector are used for the video output lines with the I/O standard set to TMDS_33.

Image Processing With Xilinx Devices

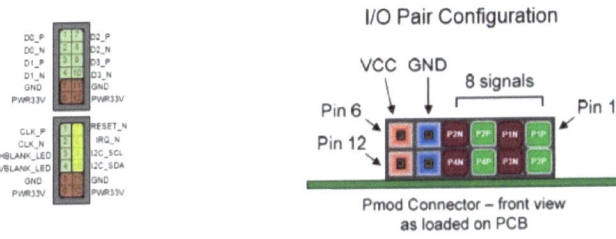

Figure 112 Pmod Pin out

With the basic hardware design in place all that remains now is to generate the software builds. Initially, I will build a bare metal application to verify that this design functions as intended. This step-by-step process stems from my strong belief in incremental verification as a project progresses.

Notes:

- You need to install the MiniZed board definition files into your Vivado /data/boards/board_files directory to work with the MiniZed dev board. If you have not already done so, they are available here.

With the hardware platform built using the Zynq-based Avnet MiniZed dev board, the next step in this adventure is to write the software so we can display images on the 7-inch touch display. To do this we need write a bare-metal software application to do the following:

- Configure the video timing controller (VTC) to generate timings required for the 800x480-pixel WGA (Wide Video Graphics Array) display.
- Create three frame buffers within the PS (processing system) DDR SDRAM.
- Configure the FLIR Lepton IR camera and store images in the current write frame buffer.
- Configure the VDMA to read from the current read frame buffer.

The first step is to configure VTC to generate video timing signals for the desired resolution. Failing to do this correctly will mean that the AXI-Stream-to-Video-Out block won't lock with the AXIS video stream.

The VTC is a core component, present in most image-processing pipelines (ISPs). The VTC's function is not just limited to generating timing signals; it also detects video input timing. This feature allows the VTC to lock its timing generation with input video streams. That's a key capability if the ISP needs to be agile and if it's to adapt on the fly to changes in input resolution.

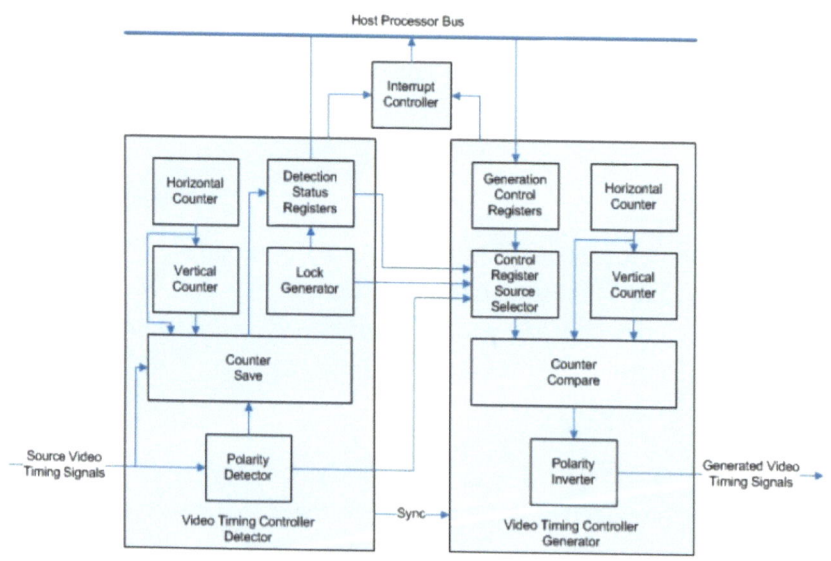

The VTC generator can be configured by either its own registers, which we update when write to those registers directly, or by the VTC detector registers. For this exercise, we need to set the VTC generator register sources correctly because we are only using the generator half of the VTC and not the detector half. The VTC's power-on default is to take configuration data from the detector registers and that's not the mode we wish to use here. To set the VTC register source, we'll use a variable of the structure type XVtc_SourceSelect in conjunction with the function XVtc_SetSource().

Image Processing With Xilinx Devices

```
memset((void *)&SourceSelect, 0, sizeof(SourceSelect));
SourceSelect.VBlankPolSrc = 1;
SourceSelect.VSyncPolSrc = 1;
SourceSelect.HBlankPolSrc = 1;
SourceSelect.HSyncPolSrc = 1;
SourceSelect.ActiveVideoPolSrc = 1;
SourceSelect.ActiveChromaPolSrc= 1;
SourceSelect.VChromaSrc = 1;
SourceSelect.VActiveSrc = 1;
SourceSelect.VBackPorchSrc = 1;
SourceSelect.VSyncSrc = 1;
SourceSelect.VFrontPorchSrc = 1;
SourceSelect.VTotalSrc = 1;
SourceSelect.HActiveSrc = 1;
SourceSelect.HBackPorchSrc = 1;
SourceSelect.HSyncSrc = 1;
SourceSelect.HFrontPorchSrc = 1;
SourceSelect.HTotalSrc = 1;
```

Together these lines of code set the VTC control-register bits 8 to 26, which determine the source for each register. Each of these bits controls a specific generator register source. For example, bit 8 controls the Frame Horizontal Size register. Setting this bit to "0" instructs the VTC to use the detector settings while a "1" instructs the VTC to use the generator's internal register settings.

Failing to do this results in writes to the detector registers having no effect on the generated video timing, which can be a rather frustrating issue to track down.

With the correct register source set, the next step is to write the timing parameters. We need the following settings for the 7-Inch touch display:

Paramter	Horizontal	Vertical
Active Video	800	480
Front Porch	40	3
Back Porch	88	35
Sync Width	128	2
Sync Polarity	0	0
Interlaced	0	0

These parameters are stored in a variable of the XVtc_Timing type. We write them into the VTC using the XVtc_SetGeneratorTiming() function:

```
vtcTiming.HActiveVideo = 800;
vtcTiming.HFrontPorch = 40;
vtcTiming.HSyncWidth = 128;
vtcTiming.HBackPorch = 88;
vtcTiming.HSyncPolarity = 0;
vtcTiming.VActiveVideo = 480;
vtcTiming.V0FrontPorch = 3; //8;
vtcTiming.V0SyncWidth = 2;
vtcTiming.V0BackPorch = 35;
vtcTiming.V1FrontPorch = 3;
vtcTiming.V1SyncWidth = 2;
vtcTiming.V1BackPorch = 35;
vtcTiming.VSyncPolarity = 0;
vtcTiming.Interlaced = 0;
```

Of course, the VDMA and the frame buffers must also be aligned with the VTC. The current design uses three frame buffers to store the output images. Each frame buffer is based on the u32 type and declared as a one-dimensional array containing the total number of pixels in the image.

The u32 type is ideal for the frame buffer because each pixel in the 7-

Image Processing With Xilinx Devices

inch touch display requires eight-bit Red, Green, and Blue values. Therefore, we need 24 bits per pixel. Each frame buffer has an associated pointer that we'll use for frame-buffer access. We initialize these pointers just after the program starts.

We use the VDMA to display the contents of the frame buffer. The key VDMA configuration parameters are stored within a variable of the type XAxiVdma_DmaSetup. It is here where we define the vertical & horizontal size, stride, and the frame-store addresses. The DMA is then configured using this data and the XAxiVdma_DmaConfig() and XAxiVdma_DmaSetBufferAddr() functions. One very important thing to remember here is that the horizontal size and stride are entered bytes. So in this example, they are set to 800 * 4 as each u32 word consists of 4 bytes.

```
vdmaDMA.FrameDelay = 0;
vdmaDMA.EnableCircularBuf = 1;
vdmaDMA.EnableSync = 0;
vdmaDMA.PointNum = 0;
vdmaDMA.EnableFrameCounter = 0;
vdmaDMA.VertSizeInput = 480;
vdmaDMA.HoriSizeInput = (800*4);
vdmaDMA.FixedFrameStoreAddr = 0;
vdmaDMA.FrameStoreStartAddr[0] = (u32)  pFrames[0];
vdmaDMA.Stride = DEMO_STRIDE;
```

We'll use code from the previous example (p1 & p2) to interface with the FLIR Lepton IR camera. This code communicates with the camera over I2C and SPI interfaces. Once the image has been received from the camera, the code copies the image into the frame buffer. However, to ensure that we use most of the available image frame, we'll use a simple digital zoom to scale up the 80x60-pixel image from the Lepton 2 camera. To do this, we output each pixel eight times to generate a 640x480-pixel display image that we'll position within the 7-inch touch display's 800x480 pixels. We set the remaining pixels to a constant color. As this is a touch display, this remaining space would be idea for command buttons and other user interfaces.

Putting all this together results in the image below. The green coloring comes from mapping the 8-bit Lepton image data into the green channel of the display.

This combination of the FLIR Lepton camera and the Zynq-based MiniZed dev board results in a very compact and cost-efficient thermal-imaging solution. The next step in our journey is to get the MiniZed's wireless communications working with PetaLinux so that we can transmit these images over the air.

EXAMPLE DESIGNS

- https://github.com/ATaylorCEngFIET/FMC-HDMI-Zed - VTC Mode detection using ADV7611
- https://github.com/ATaylorCEngFIET/UltraZed_Part18 - UltraScale+ MPSoC PL to PS Example
- https://github.com/ATaylorCEngFIET/Nexys_Video_Part220 - Direct HDMI TMDS Decoding
- https://github.com/ATaylorCEngFIET/MiniZed-FLIR-Lepton-2 - MiniZed FLIR Lepton Example
- https://github.com/ATaylorCEngFIET/FLIR_LEPTON2 - Arty Z7 FLIR Lepton Example

ABOUT THE AUTHOR

Adam Taylor is an expert in design and development of embedded systems and FPGA's for several end applications. Throughout his career, Adam has used FPGA's to implement a wide variety of solutions from RADAR to safety critical control systems, with interesting stops in image processing and cryptography along the way. Most recently he was the Chief Engineer of a Space Imaging company, being responsible for several game changing projects. Adam is the author of numerous articles on electronic design and FPGA design including over 200 blogs on how to use the Zynq for Xilinx. Adam is Chartered Engineer and Fellow of the Institute of Engineering and Technology, he is also the owner of the engineering and consultancy company Adiuvo Engineering and Training

www.ingramcontent.com/pod-product-compliance
Lightning Source LLC
Chambersburg PA
CBHW040216220526
45473CB00001B/10